Ernst Schering Research Foundation Workshop 44
Leucocyte Trafficking

Springer
Berlin
Heidelberg
New York
Hong Kong
London
Milan
Paris
Tokyo

Ernst Schering Research Foundation
Workshop 44

Leucocyte Trafficking

The Role of Fucosyltransferases and Selectins

A. Hamann, K. Asadullah, A. Schottelius,
Editors

With 43 Figures

Springer

Series Editors: G. Stock and M. Lessl

ISSN 0947-6057
ISBN 3-540-40112-1 Springer-Verlag Berlin Heidelberg New York

Library of Congress Cataloging-in-Publication Data
Leucocyte trafficking: role of fucosyltransferases and selectins / A. Hamann, K. Asadullah, A. Schottelius, editors.
    p. cm. – (Ernst Schering Research Foundation workshop; 44)
    Includes bibliographical references and index.
    ISBN 3-540-40112-1 (alk. paper)
      1. Leucocytes – Congresses. 2. Fucosyltransferases – Congresses. 3. Selectins – Congresses. I. Hamann, Alf. II. Asadullah, K. (Khusru), 1967– III. Schottelius, A. (Arndt J.G.), 1966– IV. Series.

Springer-Verlag Berlin Heidelberg New York
a member of BertelsmannSpringer Science+Business Media GmbH

http://www.springer.de

© Springer-Verlag Berlin · Heidelberg 2004
Printed in Germany

Typesetting: K+V Fotosatz GmbH, Beerfelden

21/3150/ag-5 4 3 2 1 0 – Printed on acid-free paper

Preface

Our understanding of the pathophysiology of chronic inflammatory diseases is tightly linked to understanding the sequential steps and molecular mechanisms involved in the trafficking of leukocytes from the circulation into specific tissues in health and disease. We have known for quite some time that the first step of leukocyte extravasation, the rolling of leukocytes, is primarily mediated by the interactions of selectins and their ligands. More recently there has been

The participants of the workshop

growing evidence for the crucial role that fucosyltransferases play in this process. We thus wanted to organize a workshop that would provide a forum for many highly recognized scientists to present and discuss recent developments and breakthroughs in this exciting and ever-growing research field. For this 44th ESRF Workshop on "Leukocyte Trafficking: The Role of Fucosyltransferases and Selectins" held in Berlin from 30 October to 1 November, 2002, we were successful in bringing together a group of international scientists who are leaders in their field. We are grateful for their excellent contributions to this workshop, which are summarized in these proceedings. The organizers hope that this publication will be a valuable source for scientists and clinicians alike, since this book provides an in-depth overview of the mechanisms of leukocyte trafficking and also discusses options for the pharmacological intervention to treat inflammatory diseases.

Arndt Schottelius, Khusru Asadullah, Alf Hamann

Berlin, April 2003

Contents

List of Editors and Contributors

Editors

Hamann, A.
Experimental Rheumatology, Medical Clinic,
University Hospital Charité, Schumannstr. 21/22,
10117 Berlin, Germany
e-mail: hamann@drfz.de

Schottelius, A.
Corporate Research Business, Area Dermatology, Schering AG,
13342 Berlin, Germany
e-mail: arndt.schottelius@schering.de

Asadullah, K.
Corporate Research Business Area Dermatology,
Schering AG, 13342 Berlin, Germany
e-mail: Khusru.Asadullah@schering.de

Contributors

Akdis, C.A.
Swiss Institute of Allergy and Asthma Research (SIAF),
Obere Strasse 22, 7270 Davos, Switzerland

Akdis, M.
Swiss Institute of Allergy and Asthma Research (SIAF),
Obere Strasse 22, 7270 Davos, Switzerland
e-mail: akdism@siaf.unizh.ch

Albanesi, C.
Istituto Dermopatico dell' Immacolata, IRCCS,
Via Monte di Creta 104 00167, Roma, Italy
e-mail: c.albanesi@idi.it

Alon, R.
Department of Immunology, The Weizmann Institute of Science,
Rehovot 76100, Israel
e-mail: ronalon@wicc.weizmann.ac.il

Blaser, K.
Swiss Institute of Allergy and Asthma Research (SIAF),
Obere Strasse 22, 7270 Davos, Switzerland
e-mail: blaserk@siaf.unizh.ch

Cavani, A.
Istituto Dermopatico dell' Immacolata, IRCCS,
Via Monte di Creta 104 00167, Roma, Italy
e-mail: cavani@idi.it

Ding, Y.
Department of Chemistry, University of Alberta, Edmonton,
Alberta, T6G 2G2, Canada

Dwir, O.
Department of Immunology, The Weizmann Institute of Science,
Rehovot 76100, Israel
e-mail: oren.dwir@weizmann.ac.il

Ernst, B.
Department of Chemistry, University of Alberta, Edmonton,
Alberta, T6G 2G2, Canada
e-mail: beat.ernst@unibas.ch

Girolomoni, G.
Istituto Dermopatico dell' Immacolata, IRCCS,
Via Monte di Creta 104 00167, Roma, Italy
e-mail: giro@idi.it

Grabovsky, V.
Department of Immunology, The Weizmann Institute of Science,
Rehovot 76100, Israel
e-mail: vgrabovsky@hotmail.com

Hindsgaul, O.
Department of Chemistry, University of Alberta, Edmonton,
Alberta, T6G 2G2, Canada
e-mail: de.hindsgaul@valberta.ed

Horuk, R.
Berlex, Richmond, 15049 San Pablo Avenue, Richmond,
CA 94804-0099, USA
e-mail: richard_horuk@berlex.com

Hühn, J.
Experimental Rheumatology, Medical Clinic,
University Hospital Charité, Schumannstr. 21/22,
10117 Berlin, Germany
e-mail: huehn@drfz.de

Jennrich, S.
Experimental Rheumatology, Medical Clinic,
University Hospital Charité, Schumannstr. 21/22,
10117 Berlin, Germany
e-mail: jennrich@drfz.de

Kansas, G.
Northwestern University Medical School,
Department of Microbiology-Immunology,
303 East Chicago Avenue, Chicago, IL 60611, USA
e-mail: gsk@northwestern.edu

Klein, J.
Institut für Organische Chemie,
Universität Hamburg, Martin-Luther-King-Platz 6,
20146 Hamburg, Germany

Klunker, S.
Swiss Institute of Allergy and Asthma Research (SIAF),
Obere Strasse 22, 7270 Davos, Switzerland
e-mail: klunkers@siaf.unizh.ch

Knaie, O.
Department of Chemistry, University of Alberta, Edmonton,
Alberta, T6G 2G2, Canada

Kretschmer, U.
Experimental Rheumatology, Medical Clinic,
University Hospital Charité, Schumannstr. 21/22,
10117 Berlin, Germany
e-mail: kretzschmer@drfz.de

Labbe, J.
Department of Chemistry, University of Alberta, Edmonton,
Alberta, T6G 2G2, Canada

Lühn, K.
Max-Planck-Institut für Vaskuläre Biologie,
c/o Institut für Zellbiologie, ZMBE, University Münster,
Von-Esmarch-Str. 56, 48149 Münster, Germany
e-mail: luehn@uni-muenster.de

Marquard, T.
Klinik und Poliklinik für Kinderheilkunde, Universität Münster,
Albert-Schweitzer-Str. 33, 48149 Münster, Germany
e-mail: marquard@uni-muenster.de

Mayer, M.
Institut für Organische Chemie, Universität Hamburg,
Martin-Luther-King-Platz 6, 20146 Hamburg, Germany

McEver, R.
Cardiovascular Biology Research Program,
Oklahoma Medical Research Foundation, 825 N.E. 13th Street,
Oklahoma City, OK 73104, USA
e-mail: rodger-mcever@omrf.ouhsc.edu

Meinecke, R.
Institut für Organische Chemie, Universität Hamburg,
Martin-Luther-King-Platz 6, 20146 Hamburg, Germany

Meyer, B.
Institut für Organische Chemie, Universität Hamburg,
Martin-Luther-King-Platz 6, 20146 Hamburg, Germany
e-mail: bernd.meyer@sgil.chemie.uni-hamburg.de

Möller, H.
Institut für Organische Chemie, Universität Hamburg,
Martin-Luther-King-Platz 6, 20146 Hamburg, Germany

Neffe, A.
Institut für Organische Chemie, Universität Hamburg,
Martin-Luther-King-Platz 6, 20146 Hamburg, Germany

Palcic, M.M.
Department of Chemistry, University of Alberta, Edmonton,
Alberta, T6G 2G2, Canada
e-mail: monica.palcic@ualberta.edu

Pastore, S.
Istituto Dermopatico dell' Immacolata, IRCCS,
Via Monte di Creta 104 00167, Roma, Italy
e-mail: pastore@idi.it

Schön, M.
Otto-von-Guericke-Universität, Department of Dermatology,
Leipziger Str. 44, 39120 Magdeburg, Germany
e-mail: michael.schoen@medizin.uni-magdeburg.de

Schuster, O.
Institut für Organische Chemie, Universität Hamburg,
Martin-Luther-King-Platz 6, 20146 Hamburg, Germany
e-mail: schuster@chemie.uni-hamburg.de

Syrbe, U.
Experimental Rheumatology, Medical Clinic,
University Hospital Charité, Schumannstr. 21/22,
10117 Berlin, Germany
e-mail: syrbe@drfz.de

Thiem, J.
Institut für Organische Chemie, University of Hamburg,
Martin-Luther-King-Platz 6, 20146 Hamburg, Germany
e-mail: thiem@chemie.uni-hamburg.de

Vestweber, D.
Max-Planck-Institut für Vaskuläre Biologie,
c/o Institut für Zellbiologie, ZMBE, University Münster,
Von-Esmarch-Str. 56, 48149 Münster, Germany

Wagner, B.
Institute of Molecular Pharmacy, university of Basel,
4056 Basel, Switzerland

Wild, M.
Max-Planck-Institut für Vaskuläre Biologie,
c/o Institut für Zellbiologie, ZMBE, University Münster,
Von-Esmarch-Str. 56, 48149 Münster, Germany
e-mail: wild@uni-muenster.de

Wülfken, J.
Institut für Organische Chemie, Universität Hamburg,
Martin-Luther-King-Platz 6, 20146 Hamburg, Germany

1 Selectin Ligands on T Cells

A. Hamann, U. Syrbe, U. Kretschmer, S. Jennrich, J. Hühn

1.1 Introduction

L-selectin was the first member of the selectin family identified by its function as lymphocyte homing receptor and key mediator of T cell recirculation through lymph nodes (Gallatin et al. 1983). E- and P-selectin, detected later, were not only found differentially distributed, i.e., restricted to the endothelial cell side, but also associated with inflammation rather than with homeostatic lymphocyte recirculation (Pober and Cotran 1990). A large body of data shows that both E- and P-selectin become upregulated in vitro and in vivo upon action of inflammatory mediators such as tumor necrosis factor (TNF) or others on endothelium of various tissues (Kansas 1996; McEver 1997; Vestweber 1997; Patel et al. 2002). Recruitment of myeloid cells, which express constitutively ligands for endothelial cells, is a major consequence, but early studies also reported binding of memory T cells to endothelium expressing E- or P-selectin (Shimizu et al. 1991).

The role of endothelial selectins for the recruitment of T cells received more attention when we and other groups reported that ligands for selectins are upregulated upon differentiation into effector cells and are required for their entry into inflamed tissues (Austrup et al. 1997; Wagers et al. 1998; Lim et al. 1999). These in vitro studies found a preferential expression of selectin ligands on the proinflammatory Th1 effector cell subset and suggested a link between functional polarization and expression of selectin ligands. However, data from human cells (Teraki and Picker 1997; Akdis et al. 2000) and from mouse effector cells generated in vivo (Thoma et al. 1998; Tietz et al. 1998) did not confirm these findings, as will be discussed in Sect. 1.2.2. A significant body of data demonstrated that key elements determining the absence or presence of ligands for selectins on T cells are the enzymes involved in synthesis of the sLex-related carbohydrate epitopes. Notably, mRNA expression of fucosyltransferase VII and core-2-glucosaminyltransferase has been shown to correlate with the appearance of functional ligands (Lowe et al. 1990; Knibbs et al. 1996, 1998; Wagers et al. 1997; Lim et al. 2001; Snapp et al. 2001).

1.2 Expression

1.2.1 Rapid, Transient Expression of P- and E-Selectin Ligands on CD4$^+$ T Cells In Vivo

In healthy mice, only small numbers of total T cells express considerable levels of ligands for either P- or E-selectin. For CD4$^+$ T cells, numbers are approximately 4% and 1%, respectively; during inflammatory reactions a transient increase is detected (Thoma et al. 1998; Tietz et al. 1998). P- and E-selectin binding epitopes seem to be regulated largely in parallel and differ only in quantitative terms; P-selectin ligand (P-lig)$^+$ cells are more abundant and comprise all E-ligand$^+$ cells (Tietz et al. 1998); it appears that the generation of both ligands requires different levels of fucosyltransferase activity rather than being independently regulated (Knibbs et al. 1998).

On naive T cells, selectin ligands are almost absent; however, among memory/effector cells the frequencies are much higher. Ten percent to sixty percent of P-selectin binding cells (and even more under inflammatory conditions) are found among cytokine-producing CD4$^+$ memory/effector cells (U. Kretschmer, K. Bonhagen, H.-W. Mittrücker, G. F. Debes, U. Syrbe, K.J. Erb, D. Zaiss, O. Liesenfeld, T. Kamradt, A. Hamann, manuscript submitted). In human blood, 10%–30% of memory CD4$^+$ cells express the "cutaneous lympho-cyte antigen," CLA, a carbohydrate epitope known to serve as E-se-lectin ligand and detected by the mAb HECA 452 (Berg et al. 1991; Picker et al. 1991, 1993).

Under homeostatic conditions, frequencies of P-lig-expressing cells depend on the organ harboring the effector cells: mesenteric lymph nodes, with about 10%, contain the lowest frequency of P-lig-expressing cytokine producers; peripheral (subcutaneous) lymph nodes, with 60%, contain the highest; the spleen is in between (U. Kretschmer et al., unpublished data; Campbell and Butcher 2002). Rapidly (2 days) after induction of an immune response, P-lig ex-pression was found to be upregulated on proliferating TCR-trans-genic T cells in vivo; again, the frequency of P-lig$^+$ cells was related to the organ environment and was high in subcutaneous lymph nodes but low or absent in mesenteric lymph nodes (Campbell and Butcher 2002). Whether the loss of P-lig expression on the cells re-covered from animals within 2–3 weeks after an inflammatory re-sponse (Tietz et al. 1998) is due to downregulation of the ligands or to death of the induced effector cells has not been investigated so far.

1.2.2 Lack of a Correlation Between Selectin Ligand Expression and Cytokine Phenotype In Vivo

In a systematic study we investigated the expression of P-lig on cytokine-producing effector cells found under homeostatic conditions or generated upon inflammation in different infection models and different tissues. The clear-cut result is a lack of a consistent correlation between cytokine phenotype (Th1) and P-lig expression; both interferon (IFN)-γ and interleukin (IL)-4 producing cells were equally able to express high levels of P-lig. In fact, IL-10 producing cells were in most cases the population with the highest number of P-lig$^+$ cells (U. Kretschmer et al., manuscript submitted). Similar to the in vivo activation of adoptively transferred transgenic cells by antigen plus adjuvant injection (Campbell and Butcher 2002), the frequencies of P-lig$^+$ effector/memory cells were instead related to the type of lymphoid tissue from which cells were isolated, as mentioned above, and to distinct inflammatory stimuli, as discussed in Sect. 1.2.3.

1.2.3 Expression Increases Under Inflammatory Conditions

In a skin sensitization model, increased numbers of P- and E-lig-expressing cells were found in lymph nodes draining the inflamed skin (Tietz et al. 1998). Similarly, high frequencies of both P- and E-selectin binding cells were found in mucosal and lymphoid tissues in severe combined immunodeficiency (SCID) colitis (Thoma et al. 1998). This also applied to the mesenteric lymph node, where, under homeostatic conditions or upon immunization, low frequencies were found (Campbell and Butcher 2002). Thus, strong inflammatory signals are able to overrule the organ-associated differences in selectin ligand expression. Accordingly, in infection models we found high frequencies of P-lig-expressing effector cells in inflamed tissues such as lung or liver, whereas effector T cells from other noninflamed tissues of the infected animals, including lymphoid tissues, did not display significantly increased frequencies. This suggests that the inflammatory environment leads to the upregulation of selectin ligands or supports a preferential accumulation of selectin ligand-positive cells.

Interestingly, upregulation of P-lig expression is confined to the dominating type of effector cells, either Th1 cells in Th1-dominated or Th2 cells in Th2-dominated infection models, respectively: in a parasite-induced inflammation, the IL-4-producing T cells were those with the highest frequency of P-lig expression, whereas in viral infection IFN-γ producers displayed a high frequency (U. Kretschmer et al., manuscript submitted). This suggests that upregulation of P-lig is restricted to the recently activated effector cells, and might support their recirculation into the inflamed tissue.

1.3 Regulation

1.3.1 Inducing Factors for Selectin Ligands on CD4[+] T Cells

The findings mentioned in the previous section suggest a high degree of selectivity in the factors inducing the expression of selectin ligands on T cells in vivo. Although some work has been done to elucidate factors regulating the expression of fucosyltransferase VII and functional selectin binding epitopes, major issues in the regulation of these inflammation-specific homing mechanisms are still unknown. Initial studies pointed to a key role of IL-12 in induction of fucosyltransferase and functional selectin ligands. In in vitro cultures, IL-12 is the key factor leading to both differentiation into the Th1 subset and upregulation of selectin ligand expression (Leung et al. 1995; Austrup et al. 1997; Wagers et al. 1998; Lim et al. 1999). In the absence of IL-12 or by use of T cells from STAT-4-deficient mice, which is the main signal transducing molecule for IL-12, the activated cells acquire lower, but still significant levels of selectin ligand expression; natural memory/effector cells from STAT-4$^{-/-}$ mice even displayed similar frequencies of P-lig as cells from wild-type animals (White et al. 2001; U. Syrbe and U. Kretschmer, unpublished data). This indicates that IL-12 is able to induce selectin ligand expression, but is not indispensable. Moreover, these and further data on the regulation of fucosyltransferase VII in human T cells suggest that either IL-12 might act independently from STAT-4

(White et al. 2001), and/or that for the generation of P-lig-expressing cells in vivo IL-12 is not of major importance. In the human,
transforming growth factor (TGF)-β, but not a variety of other cytokines, proved to upregulate selectin ligand expression on T cells in
addition to and in a synergistic way with IL-12 (Picker et al. 1993;
Wagers and Kansas 2000). In the mouse, TGF-β seems not to be
able to induce selectin ligands (U. Syrbe et al., unpublished data).

Whether distinct factors exert an inhibitory effect on selectin ligand expression is unclear so far. Some papers claimed a suppressive effect of IL-4 on selectin ligand expression (Wagers et al. 1998;
Lim et al. 1999); we could not find a major, direct role of IL-4 and
instead assume indirect effects on T cell activation or survival.

In vivo and in vitro, induction of selectin ligand expression was
only observed on T cells upon activation and required cell proliferation (Campbell and Butcher 2002; Syrbe et al.). This led us to study
the role of T cell receptor (TCR) signaling and proliferation for the
generation of selectin ligands in more detail. First, by use of a transgenic TCR model, we found upregulation of selectin ligands restricted to cells receiving a signal through the TCR; soluble factors
present in Th1 cultures, where most cells reacted with activation and
ligand induction, were not sufficient to induce expression on cells
lacking the respective TCR (U. Syrbe et al., unpublished data). Similarly, suppression of TCR downstream signaling by cyclosporin A
was found to prevent P-lig induction.

Moreover, induction of selectin ligands in naive T cells was largely restricted to cells completing one or more cell cycles, similar to
findings in vivo (Campbell and Butcher 2002). Interestingly, this did
not apply to cells of memory phenotype. In this population, a certain
percentage (about 10%) already expresses P-lig without activation; a
major fraction of the memory cells can upregulate P-lig expression
upon stimulation without need for cell cycling, indicating that differentiation into effector/memory cells is linked with permanent alterations in the capacity to express these homing receptors.

1.4 Function

1.4.1 Role for Migration of Effector T Cells into the Inflamed Skin and Redundant Role of E- and P-Selectins

When the small fraction of effector/memory cells expressing selectin ligands in vivo is isolated and reinjected into mice, a very high number migrates into the inflamed skin (Tietz et al. 1998). After 24 h, up to 20% of E-selectin binding cells accumulate there, in contrast to very low numbers of total CD4$^+$ cells (Y. Allemand et al., unpublished data). This accumulation can be blocked completely by antibodies against P- and E-selectin, provided they are given in combination. Blocking either pathway alone results in less than 50% inhibition, indicating that both selectins synergize in recruiting effector/memory cells into the inflamed skin and act in a largely redundant way (Tietz et al. 1998). Similarly, entry of in vitro generated (Th1-) effector cells into the inflamed skin (which is less efficient than that of ex vivo isolated cells) is completely dependent on selectins (Austrup et al. 1997). Accordingly, Th1 cells from fucosyltransferase VII-deficient mice are unable to enter inflamed skin (Erdmann et al. 2002). A role of endothelial selectins in homing to the inflamed skin has also been demonstrated in a primate model (Silber et al. 1994).

A number of studies found a preferential expression of the CLA-epitope on T cells residing in skin, but not other organs (Picker et al. 1990). Moreover, memory cells specific for cutaneous antigens, but not those generated in systemic atopies were found to be CLA-positive (Abernathy-Carver et al. 1995; Santamaria-Babi et al. 1995). The conclusion was that the E-selectin ligand CLA serves as skin-specific homing receptor (Berg et al. 1991).

However, at least in mouse models, expression of selectin ligands and function as homing receptor are not confined to the skin. As mentioned above, high frequencies of P- and E-selectin binding T effector cells were found in the inflamed gut as well as in inflamed lungs (Thoma et al. 1998; U. Kretschmer et al., unpublished data). Blocking both selectins together or use of fucosyltransferase VII-deficient T cells resulted in almost complete abrogation of homing into inflamed joints in arthritis models (Austrup et al. 1997) and considerably reduced accumulation in the inflamed colon or in the inflamed lung (Wolber et al. 1997, 1998; U. Kretschmer et al., unpublished data).

In conclusion, E- and P-selectin act together, in a largely redundant way, in recruiting T effector/memory cells into various inflamed tissues, whereby their role in the inflamed skin seems to be especially dominant. Accordingly, E- and P-selectin represent predominantly an inflammation-specific rather than an organ-specific homing mechanism.

1.4.2 Protective and Adverse Effects of Selectin Targeting in Disease Models

In a variety of models such as skin inflammation, arthritis, or allergic airway responses, blocking of endothelial selectins or deletion of selectins or fucosyltransferase genes has resulted in prevention or amelioration of inflammatory symptoms (Tipping et al. 1994; Staite et al. 1996; De Sanctis et al. 1997; Broide et al. 1998; Issekutz et al. 2001; Smithson et al. 2001; Lukacs et al. 2002).

However, in a few models such as trinitrobenzene (TNBS)-induced colitis, glomerulonephritis, or arthritis, aggravation of disease was reported upon deletion or blocking of selectins (Bullard et al. 1999; McCafferty et al. 1999; Rosenkranz et al. 1999). Furthermore, spontaneous development of dermal or pulmonary inflammations and increased infections have been reported in some P/E-selectin-deficient mice strains (Frenette et al. 1996; Collins et al. 2001; Forlow et al. 2002). Although in some of these studies, the genetic background of the animal was not carefully controlled and might have contributed to conflicting results, it can not be excluded that blocking of selectins is not always beneficial. Three mechanisms can be imagined by which adverse effects could occur:

1. Blockade of selectins and concomitant inhibition of leukocyte accumulation could lead to uncontrolled infection, e.g., in the colitis models.
2. It is not known whether regulatory T cells require homing mechanisms and must enter inflamed sites to control immune reactions and prevent autoimmunity. If this is the case, prolonged blockade

of selectins could disable the resolution of an inflammatory response by accumulation of suppressive cell populations.
3. Alterations in hemopoiesis observed in E-/P-selectin knock-out animals (Frenette et al. 1996) could indicate that this pathway also has a role in homeostatic functions.

1.4.3 Functions Beyond Homing

It has been shown that triggering of PSGL-1, the main ligand for P- as well as E-selectin (Fuhlbrigge et al. 1997), leads to intracellular signals, including tyrosine phosphorylation, e.g., of syk, and SRE-dependent transcriptional activation (Hidari et al. 1997; Urzainqui et al. 2002). Moreover, activation of cytokine production was found in T cells and other leukocytes upon engagement of PSGL-1 (Damle et al. 1992; Weyrich et al. 1995; J. Hühn et al., unpublished data). This suggests that the encounter of T effector cells with inflamed endothelium or with activated platelets expressing P-selectin might sensitize T cells and leukocytes for subsequent generation of effector functions.

1.5 Open Questions

A variety of studies support the importance of selectin-mediated mechanisms in inflammatory reactions. With regard to T cells, selectin-dependent recruitment into inflamed sites has been clearly demonstrated. However, its actual impact on the pathophysiology is less clear, as in the knock-out models so far available where both leukocyte and T effector cell recruitment are affected. Consequently, suppression of disease symptoms in these models could rely on either leukocyte or lymphocyte blockade, or both.

Another issue as yet unresolved is the role of selectin-mediated migration in the function of regulatory T cells, which have important roles in prevention and control of autoimmune reactions. Thus, adverse effects observed in some situations after deletion of E-/P-selectins could be due to an undesirable modulation of the localization of regulatory T cells. The recent identification of different subsets of regulatory T cells (Lehmann et al. 2002) and upcoming studies on their function in inflammatory diseases might help to clarify this point.

References

Abernathy-Carver KJ, Sampson HA, Picker LJ, Leung DY (1995) Milk-induced eczema is associated with the expansion of T cells expressing cutaneous lymphocyte antigen. J Clin Invest 95:913–918

Akdis M, Klunker S, Schliz M, Blaser K, Akdis CA (2000) Expression of cutaneous lymphocyte-associated antigen on human CD4(+) and CD8(+) Th2 cells. Eur J Immunol 30:3533–3541

Austrup F, Vestweber D, Borges E, Löhning M, Bräuer R, Herz U, Renz H, Hallmann R, Scheffold A, Radbruch A, Hamann A (1997) P-and E-selectin mediate recruitment of T helper 1 but not T helper 2 cells into inflamed tissues. Nature 385:81–83

Berg EL, Yoshino T, Rott LS, Robinson MK, Warnock RA, Kishimoto TK, Picker LJ, Butcher EC (1991) The cutaneous lymphocyte antigen is a skin lymphocyte homing receptor for the vascular lectin endothelial cell-leukocyte adhesion molecule 1. J. Exp. Med. 174:1461–1466

Broide DH, Sullivan S, Gifford T, Sriramarao P (1998) Inhibition of pulmonary eosinophilia in P-selectin- and ICAM-1-deficient mice. Am J Respir Cell Mol Biol 18:218–225

Bullard DC, Mobley JM, Justen JM, Sly LM, Chosay JG, Dunn CJ, Lindsey JR, Beaudet AL, Staite ND (1999) Acceleration and increased severity of collagen-induced arthritis in P-selectin mutant mice. J Immunol 163:2844–2849

Campbell DJ, Butcher EC (2002) Rapid acquisition of tissue-specific homing phenotypes by CD4+ T cells activated in cutaneous or mucosal lymphoid tissues. J Exp Med 195:135–141

Collins RG, Jung U, Ramirez M, Bullard DC, Hicks MJ, Smith CW, Ley K, Beaudet AL (2001) Dermal and pulmonary inflammatory disease in E-selectin and P-selectin double-null mice is reduced in triple-selectin-null mice. Blood 98:727–735

Damle NK, Klussman K, Dietsch MT, Mohagheghpour N, Aruffo A (1992) GMP-140 (P-selectin/CD62) binds to chronically stimulated but not resting CD4+ T lymphocytes and regulates their production of proinflammatory cytokines. Eur J Immunol 22:1789–1793

De Sanctis GT, Wolyniec WW, Green FH, Qin S, Jiao A, Finn PW, Noonan T, Joetham AA, Gelfand E, Doerschuk CM, Drazen JM (1997) Reduction of allergic airway responses in P-selectin-deficient mice. J Appl Physiol 83:681–687

Erdmann I, Scheidegger EP, Koch FK, Heinzerling L, Odermatt B, Burg G, Lowe JB, Kundig TM (2002) Fucosyltransferase VII-deficient mice with defective E-, P-, and L- selectin ligands show impaired CD4$^+$ and CD8$^+$ T cell migration into the skin, but normal extravasation into visceral organs. J Immunol 168:2139–2146

Forlow SB, White EJ, Thomas KL, Bagby GJ, Foley PL, Ley K (2002) T cell requirement for development of chronic ulcerative dermatitis in E- and P-selectin-deficient mice. J Immunol 169:4797–4804

Frenette PS, Mayadas TN, Rayburn H, Hynes RO, Wagner DD (1996) Susceptibility to infection and altered hematopoiesis in mice deficient in both P- and E-selectins. Cell 84:563–574

Fuhlbrigge RC, Kieffer JD, Armerding D, Kupper TS (1997) Cutaneous lymphocyte antigen is a specialized form of PSGL-1 expressed on skin homing T cells. Nature 389:978–981

Gallatin WM, Weissman IL, Butcher EC (1983) A cell-surface molecule involved in organ-specific homing of lymphocytes. Nature 304:30–34

Hidari KI, Weyrich AS, Zimmerman GA, McEver RP (1997) Engagement of P-selectin glycoprotein ligand-1 enhances tyrosine phosphorylation and activates mitogen-activated protein kinases in human neutrophils. J Biol Chem 272:28750–28756

Issekutz AC, Mu JY, Liu G, Melrose J, Berg EL (2001) E-selectin, but not P-selectin, is required for development of adjuvant-induced arthritis in the rat. Arthritis Rheum 44:1428–1437

Kansas GS (1996) Selectins and their ligands: current concepts and controversies. Blood 88:3259–3287

Knibbs RN, Craig RA, Maly P, Smith PL, Wolber FM, Faulkner NE, Lowe JB, Stoolman LM (1998) Alpha(1,3)-fucosyltransferase VII-dependent synthesis of P- and E-selectin ligands on cultured T lymphoblasts. J Immunol 161:6305–6315

Knibbs RN, Craig RA, Natsuka S, Chang A, Cameron M, Lowe JB, Stoolman LM (1996) The fucosyltransferase FucT-VII regulates E-selectin ligand synthesis in human T cells. J Cell Biol 133:911–920

Lehmann J, Huehn J, de la Rosa M, Maszyna F, Kretschmer U, Brunner M, Scheffold A, Krenn V, Hamann A (2002) Expression of the integrin alphaEbeta7 identifies unique subsets of CD25$^+$ as well as CD25- regulatory T cells. Proceed Natl Acad Sci 99:13031–13036

Leung DY, Gately M, Trumble A, Ferguson Darnell B, Schlievert PM, Picker LJ (1995) Bacterial superantigens induce T cel l expression of the skin-selective homing receptor, the cutaneous lymphocyte-associated antigen, via stimulation of interleukin 12 production. J. Exp. Med. 181:747–753

Lim Y, Xie H, Come C, Alexander S, Grusby M, Lichtman A, Luscinskas F (2001) IL-12, STAT4-dependent upregulation of CD4($^+$) T cell core 2 beta-1,6-n-acetylglucosaminyltransferase, an enzyme essential for biosynthesis of P-selectin ligands. J Immunol 167:4476–4484

Lim YC, Henault L, Wagers AJ, Kansas GS, Luscinskas FW, Lichtman AH
(1999) Expression of functional selectin ligands on Th cells is differentially regulated by IL-12 and IL-4. J Immunol 162:3193–3201

Lowe JB, Stoolman LM, Nair RP, Larsen RD, Berhend TL, Marks RM
(1990) ELAM-1–dependent cell adhesion to vascular endothelium determined by a transfected human fucosyltransferase cDNA. Cell 63:475–484

Lukacs NW, John A, Berlin A, Bullard DC, Knibbs R, Stoolman LM (2002)
E- and P-selectins are essential for the development of cockroach aller-
gen-induced airway responses. J Immunol 169:2120–2125

McCafferty DM, Smith CW, Granger DN, Kubes P (1999) Intestinal inflammation in adhesion molecule-deficient mice: an assessment of P-selectin
alone and in combination with ICAM-1 or E-selectin. J Leukoc Biol
66:67–74

McEver RP (1997) Regulation of expression of E-selectin and P-selectin. In
Vestweber D (ed). The selectins: initiators of leucocyte endothelial adhesion. Harwood Academic Publishers, Amsterdam, pp 31–48

Patel KD, Cuvelier SL, Wiehler S (2002) Selectins: critical mediators of leukocyte recruitment. Semin Immunol 14:73–81

Picker LJ, Kishimoto TK, Smith CW, Butcher EC (1991) ELAM-1 is an adhesion molecule for skin-homing T-cells. Nature 349:796–799

Picker, LJ, SA Michie, LS Rott EC Butcher (1990) A unique phenotype of
skin-associated lymphocytes in humans. Preferential expression of the
HECA-452 epitope by benign and malignant T cells at cutaneous sites.
Am. J. Pathol. 136:1053–1068

Picker LJ, Treer JR, Ferguson DB, Collins PA, Bergstresser PR, Terstappen
LW (1993) Control of lymphocyte recirculation in man. II. Differential
regulation of the cutaneous lymphocyte-associated antigen, a tissue-selective homing receptor for skin-homing T cells. J Immunol 150:1122–1136

Pober JS, Cotran RS (1990) Cytokines and endothelial cell biology. Physiol
Rev 70:427–451

Rosenkranz AR, Mendrick DL, Cotran RS, Mayadas TN (1999) P-selectin
deficiency exacerbates experimental glomerulonephritis: a protective role
for endothelial P-selectin in inflammation. J Clin Invest 103:649–659

Santamaria-Babi LF, Moser R, Perez-Soler MT, Picker LJ, Blaser K, Hauser
C (1995) Migration of skin-homing T cells across cytokine-activated human endothelial cell layers involves interaction of the cutaneous lymphocyte-associated antigen (CLA), the very late antigen-4 (VLA-4), and the
lymphocytge function-associated antigen-1 (LFA-1) J Immunol
154:1543–1550

Shimizu Y, Shaw S, Graber N, Gopal TV, Horgan KJ, Van Seventer G, Newman W (1991) Activation-independent binding of human memory T cells
to adhesion molecule ELAM-1 [see comments]. Nature 349:799–802

Silber A, Newman W, Sasseville VG, Pauley D, Beall D, Walsh DG, Ring-
ler DJ (1994) Recruitment of lymphocytes during cutaneous delayed hy-

persensitivity in nonhuman primates is dependent on E-selectin and vascular cell adhesion molecule 1. J Clin Invest 93:1554–1563

Smithson G, Rogers CE, Smith PL, Scheidegger EP, Petryniak B, Myers JT, Kim DS, Homeister JW, Lowe JB (2001) Fuc-TVII is required for T helper 1 and T cytotoxic 1 lymphocyte selectin ligand expression and recruitment in inflammation, and together with Fuc-TIV regulates naive T cell trafficking to lymph nodes. J Exp Med 194:601–614

Snapp KR, Heitzig CE, Ellies LG, Marth JD, Kansas GS (2001) Differential requirements for the O-linked branching enzyme core 2 beta1–6-N-glucosaminyltransferase in biosynthesis of ligands for E-selectin and P-selectin. Blood 97:3806–3811

Staite ND, Justen JM, Sly LM, Beaudet AL, Bullard DC (1996) Inhibition of delayed-type contact hypersensitivity in mice deficient in both E-selectin and P-selectin. Blood 88:2973–2979

Teraki Y, Picker LJ (1997) Independent regulation of cutaneous lymphocyte-associated antigen expression and cytokine synthesis phenotype during human CD4$^+$ memory T cell differentiation. J Immunol 159:6018–6029

Thoma S, Bonhagen K, Vestweber D, Hamann A, Reimann J (1998) Expression of selectin-binding epitopes and cytokines by CD4$^+$ T cells repopulating scid mice with colitis. Eur J Immunol 28:1785–1797

Tietz W, Allemand Y, Borges E, von Laer D, Hallmann R, Vestweber D, Hamann A (1998) CD4$^+$ T-cells only migrate into inflamed skin if they express ligands for E- and P-selectin. J. Immunol. 161:963–970

Tipping PG, Huang XR, Berndt MC, Holdsworth SR (1994) A role for P selectin in complement-independent neutrophil-mediated glomerular injury. Kidney Int 46:79–88

Urzainqui A, Serrador JM, Viedma F, Yanez-Mo M, Rodriguez A, Corbi AL, Alonso-Lebrero JL, Luque A, Deckert M, Vazquez J, Sanchez-Madrid F (2002) ITAM-based interaction of ERM proteins with Syk mediates signaling by the leukocyte adhesion receptor PSGL-1. Immunity 17:401–412

Vestweber D (1997) The selectins: initiators of leucocyte endothelial adhesion, Harwood Academic Publishers, Amsterdam.

Wagers AJ, Kansas GS (2000) Potent induction of alpha(1,3)-fucosyltransferase VII in activated CD4$^+$ T cells by TGF-beta 1 through a p38 mitogen-activated protein kinase-dependent pathway. J Immunol 165:5011–5016

Wagers AJ, Stoolman LM, Kannagi R, Craig R, Kansas GS (1997) Expression of leukocyte fucosyltransferases regulates binding to E-selectin: relationship to previously implicated carbohydrate epitopes. J Immunol 159:1917–1929

Wagers AJ, Waters CM, Stoolman LM, Kansas GS (1998) Interleukin 12 and interleukin 4 control T cell adhesion to endothelial selectins through opposite effects on alpha1, 3-fucosyltransferase VII gene expression. J Exp Med 188:2225–2231

Weyrich AS, McIntyre TM, McEver RP, Prescott SM, Zimmerman GA (1995) Monocyte tethering by P-selectin regulates monocyte chemotactic protein-1 and tumor necrosis factor-alpha secretion. Signal integration and NF-kappa B translocation [see comments]. J Clin Invest 95:2297–2303

White SJ, Underhill GH, Kaplan MH, Kansas GS (2001) Cutting edge: differential requirements for Stat4 in expression of glycosyltransferases responsible for selectin ligand formation in Th1 cells. J Immunol 167:628–631

Wolber FM, Curtis JL, Maly P, Kelly RJ, Smith P, Yednock TA, Lowe JB, Stoolman LM (1998) Endothelial selectins and alpha4 integrins regulate independent pathways of T lymphocyte recruitment in the pulmonary immune response. J Immunol 161:4396–4403

Wolber FM, Curtis JL, Milik AM, Fields T, Seitzman GD, Kim K, Kim S, Sonstein J, Stoolman LM (1997) Lymphocyte recruitment and the kinetics of adhesion receptor expression during the pulmonary immune response to particulate antigen. Am J Pathol 151:1715–1727

2 Skin Homing T Cells

M. Akdis, S. Klunker, K. Blaser, C. A. Akdis

2.1 Skin Homing of T Cells

It has been proposed that differential organ-specific trafficking of CD4$^+$ Th1 and Th2 cells promote different inflammatory reactions. Skin represents a functionally distinct immune compartment, and chronic inflammation of the skin is generally associated with tissue infiltration by T cells (Akdis et al. 2000a; Bos and Kapsenberg 1993; Leung et al. 1983). The great majority of these T cells homing to the skin are of the CD45RO$^+$ memory/effector phenotype and express the skin-selective homing receptor, cutaneous lymphocyte-associated antigen (CLA; Picker et al. 1990). The CLA epitope consist of a sialyl-Lewis^x carbohydrate and corresponds to a posttranslational modification of the P-selectin glycoprotein ligand 1 (PSGL-1; Fuhlbrigge et al. 1997; Sasaki et al. 1994). It is characterized by specific binding to the monoclonal antibody (mAb) HECA-452 (Picker et al.

1990). CLA binds to its vascular counter receptor, E-selectin (CD62E), which is expressed on inflamed superficial dermal postcapillary venules and endothelial cells (Picker et al. 1991; Rossiter et al. 1994). CLA$^+$ CD45RO$^+$ T cells migrate across activated endothelium using CLA/E-selectin, VLA-4/VCAM-1, and LFA-1/ICAM-1 interactions (Santamaria Babi et al. 1995a). The generation of CLA on T cells undergoing naive to memory transition in skin-draining lymph nodes (Picker et al. 1993) requires α (1,3)-fucosyltransferase (FucT-VII) activity (Fuhlbrigge et al. 1997; Sasaki et al. 1994). Thus, CLA expression predominantly reflects the regulated activity of the glycosyltransferase, FucT-VII. CLA is upregulated by interleukin (IL)-12, which also enhances FucT-VII expression (Blander et al. 1999; Leung et al. 1995; Lim et al. 1999; Wagers et al. 1998).

Induction of CLA expression by superantigens may play an important role in the pathogenesis of disorders associated with superantigen-producing staphylococci such as atopic dermatitis (AD; Herz et al. 1998; Leung et al. 1995; Leyden et al. 1974). Staphylococcal superantigens secreted at the skin surface may penetrate through the inflamed skin and stimulate epidermal macrophages or Langerhans cells to produce IL-1, tumor necrosis factor (TNF), and IL-12. Superantigen-stimulated Langerhans cells may migrate to skin-associated lymph nodes, and serve as antigen-presenting cells (APC). They can upregulate the expression of CLA by IL-12 production (Rook et al. 1994) and influence the functional profile of virgin T cells. Moreover, superantigens presented by keratinocytes, Langerhans cells, and macrophages can stimulate T cells in the skin, and this second round of stimulation can induce CLA formation (Lim et al. 1999). Local production of IL-1 and TNF may induce expression of E-selectin on vascular endothelium (Leung et al. 1991), allowing an initial migration of CLA$^+$ memory/effector cells. IL-12 increases CLA expression on both T cell subsets activated by antigens or superantigens, thereby, increasing their efficiency of recirculation to the skin. Together, these mechanisms tend to markedly amplify the initial cutaneous inflammation. Moreover, inflamed skin may favor the progression of the staphylococcal skin colonization.

In addition, CLA is expressed by the malignant T cells of chronic-phase cutaneous T cell lymphoma (mycosis fungoides and Sezary syndrome), but not by nonskin-associated T cell lymphomas

(Heald et al. 1993; Picker et al. 1990). CLA is expressed on less than 10% of liver-infiltrating lymphocytes of acute allograft rejection and primary biliary cirrhosis patients, although E-selectin is highly expressed on endothelium (Adams et al. 1996). In addition, CLA^+ T cells were enriched on skin-infiltrating lymphocytes but not on lymphocytes in the joints of psoriatic arthritis (Jones et al. 1997). In AD, circulating allergen-specific memory/effector T cells expressing CLA have been demonstrated to be activated and regulate IgE by secretion of an IL-13-dominated cytokine pattern and delay eosinophil apoptosis by IL-5 (Akdis et al. 1997, 1999a; Santamaria Babi et al. 1995b). Studies focused on intralesional cytokine patterns of mostly CLA expressing skin-infiltrating T cells in AD demonstrated higher interferon (IFN)-γ and less IL-4, but still high IL-5 and IL-13 production (Akdis et al. 1999a; Grewe et al. 1994; Hamid et al. 1994; Thepen et al. 1996).

Recent studies in mice suggest that adoptively transferred Th1 cells are preferentially recruited to cutaneous DTH reactions compared with Th2 cells (Austrup et al. 1997). In addition, in vitro differentiated Th1 but not Th2 cells have been shown to bind E-selectin, and expression of functional selectin ligands are upregulated by IL-12 and inhibited by IL-4 by opposite effects on FucT-VII gene expression in mice (Blander et al. 1999; Lim et al. 1999; Wagers et al. 1998). CLA expression is regulated by the same mechanisms in both $CD4^+$ and $CD8^+$ T cells. The CLA molecule was expressed on Th1 cells during the differentiation process as previously shown (Blander et al. 1999; Lim et al. 1999; Wagers et al. 1998). More importantly, CLA can be induced on Th2 cells by T cell stimulation with bacterial superantigen and/or IL-12 challenge. These Th2 cells demonstrated the same cytokine profile as the T cells found in skin biopsies and in peripheral blood of AD patients (Akdis et al. 2000b).

2.2 Expression of Cutaneous Lymphocyte-Associated Antigen on Human T Cells

Infection or other damage induces the local production of distinct cytokines by tissue cells and antigen-presenting cells, initiating the differentiation of T cells reacting to the antigen into either type 1 or type 2 cells (Mosmann and Sad 1996). IL-12 drives naive T cell differentiation toward type 1 phenotype and IL-4 drives toward type 2 (Mosmann and Sad 1996; Sallusto et al. 1997). CD4$^+$ Th1 cells are involved in cell-mediated inflammatory reactions. Their cytokines activate cytotoxic and inflammatory functions and induce delayed-type hypersensitivity reactions. Th2 cytokines support antibody production, particularly IgE responses, eosinophil differentiation and function, and associated allergic responses (Mosmann and Sad 1996). There is now clear evidence for heterogeneity of CD8$^+$ T cell functions. CD8$^+$ T cells may not act solely as effector cells concerning the elimination of viral and other intracellular pathogens (Tc1). They can secrete Th2 cytokines and help B cells for Ab production (Tc2; Conlon et al. 1995; Kemeny et al. 1994). In allergic inflammations of the skin, a considerable amount of CD8$^+$ T cells, in addition to CD4$^+$ T cells, were found to infiltrate skin, suggesting an important role for T cells of both subsets (Akdis et al. 1999 a, b).

Accordingly, the regulation of CLA on primed human Th1 and Th2 cells in CD4$^+$, and Tc1 and Tc2 cells in CD8$^+$ subsets has been investigated (Akdis et al. 2000b). Purified CD45RA$^+$, CD4$^+$, and CD8$^+$ T cells were cultured with IL-2 in the presence of IL-12 or IL-4. IL-12, but not IL-4, induced CLA expression on both CD4$^+$ and CD8$^+$ T cells. Consequently, after differentiation, Th1 and Tc1 cells expressed CLA, whereas Th2 and Tc2 cells did not express CLA on their surface. Anti-CD3 stimulation in the absence of serum in the culture medium was sufficient to induce CLA on Th2 cells. We further investigated factors that regulate CLA expression in serum-containing medium. IL-4 inhibited CLA and related a-fucosyltransferase mRNA expression. IL-12 and/or staphylococcal enterotoxin B (SEB) stimulation upregulated CLA expression on either Th2 and Tc2 cells of CD4$^+$ or CD8$^+$ subsets. Stimulation of the cells with SEB in the presence of autologous, irradiated peripheral blood mononuclear cells (PBMC) induced CLA expression on both Th1 and Th2 cells. Neutralization of IL-12 in these cultures significantly downregulated the surface CLA expression on both Th1 and Th2 cells, demonstrating that superantigen-induced IL-12 plays a major role on the induction of skin-selective homing ligand (Akdis et al. 2000b).

We also analyzed whether T cells show any limitation in the expression of skin-selective homing ligand in continuous cultures of

CD45RO$^+$ T cells (Akdis et al. 2000b). For this purpose, CLA$^+$ and CLA$^-$ subsets of CD45RO$^+$, CD4$^+$, and CD8$^+$ T cells were purified from peripheral blood. The cells were incubated for resting for 14 days in cultures containing low amounts of IL-2. CLA was down-regulated on resting T cells within 2 weeks. Subsequently, all four T cell subsets were restimulated with anti-CD2, anti-CD3, and anti-CD28 mAbs in the presence of IL-2 and IL-12. CLA was highly induced on both CLA$^+$ and CLA$^-$ cells after 7 days. The cells were rested again for additional 14 days. CLA was downregulated a second time on all subsets. These experiments demonstrate that there is no restriction for CLA expression in T cell subsets. CLA is down-regulated on resting T cells and can be induced repeatedly on CLA$^-$ T cells.

We further analyzed whether there is a limitation of CLA expression on human T cells by using nonskin-related, antigen-specific T cell clones. Regulation of CLA expression by cytokines was investigated in bee venom phospholipase A$_2$-specific T cell clones. Five different T cell clones of Th1, Th2, Th0, and Tr1 phenotypes were analyzed for CLA expression. Cells were stimulated with the phospholipase A$_2$ antigen in the presence of autologous irradiated PBMC as APC and IL-2. The addition of IL-4 to cultures significantly decreased CLA expression in T cell clones of Th2, Th0, and Tr1 phenotypes. There was no significant effect on Th1 clones. IL-12 enhanced CLA expression in all five clones, but to a lesser extent in the Th2 clone.

In conclusion, these studies demonstrate that the expression of skin homing ligand differs in T cell populations after they have differentiated from naive T cells. Apparently, this is a consequence of the regulatory influences by exogenous cytokines and superantigens on those T cell subsets. There was no principle limitation for CLA expression on T cells. CLA can be induced on Th2 and Tc2 cells, on CLA$^-$ T cells, and on nonskin-related antigen-specific T cell clones by IL-12. T cell stimulation via T cell receptor was sufficient; however, it was strictly controlled by serum factors. IL-12 responsiveness of Th2 cells was an important permissive factor for CLA expression in the presence of serum.

2.3 T Cell-Mediated Effector Mechanisms in Atopic Dermatitis

2.3.1 The Role of IL-5 and IL-13 in Atopic Dermatitis

Although most patients with atopic dermatitis (AD) show high concentrations of total and allergen-specific IgE in blood and skin, some of them express normal IgE levels and show no allergen-specific IgE antibodies. The diagnostic criteria of AD by Hanifin and Rajka (Hanifin and Rajka 1980) can be fulfilled also in the absence of elevated total IgE and specific IgE to food or environmental allergens. This suggests that elevated IgE levels and IgE sensitization are not prerequisites in the pathogenesis of the disease. The subgroup of AD patients with normal IgE levels and without specific IgE sensitization has been termed the nonallergic form of AD (NAD), nonatopic eczema, non-AD, or intrinsic-type AD (Akdis et al. 1999a; Wüthrich 1978). Recent data suggest that T cells are likely involved in the pathogenesis of AD and NAD. $CD4^+$ and $CD8^+$ subsets of skin-infiltrating T cells, as well as skin-homing CLA^+ T cells from peripheral blood, equally responded to the superantigen, SEB, and produce IL-2, IL-5, IL-13, and IFN-γ in both forms of the disease (Akdis et al. 1999a; Akdis et al. 1999b). Interestingly, skin T cells from AD patients express higher IL-5 and IL-13 levels compared to NAD patients. Thus, T cells isolated from skin biopsies of AD, but not from NAD patients, induced high IgE production in cocultures with normal B cells, which is mediated by IL-13. In addition, B cell activation with high CD23 expression is observed in the peripheral blood of AD, but not NAD patients (Akdis et al. 1999a). These findings suggest a lack of IL-13-induced B cell activation and consequent IgE production in nonatopic eczema, although high numbers of T cells are present in lesional skin of both types (Akdis et al. 1999a). More importantly, IL-4 and IL-13 neutralization in B cell cocultures with peripheral blood CLA^+ skin-homing T cells or skin-infiltrating T cells demonstrated that IL-13 represents the major cytokine for induction of hyper-IgE production in AD (Akdis et al. 1999a,b 1997).

Cytokine determinations from peripheral blood CLA^+ T cells and skin biopsies of AD patients show increased IL-5 expression (Akdis 1999a,b). Accordingly, supernatants from CLA^+ T cells of both

CD4$^+$ and CD8$^+$ subsets extend the life span of freshly purified eosinophils in vitro, whereas supernatants of CLA$^-$ T cells do not influence eosinophil survival. Neutralization of cytokines demonstrated the predominant role of IL-5 secreted from CLA$^+$-T cells in prolonged eosinophil survival in AD (Akdis et al. 1999b).

2.3.2 Keratinocyte Apoptosis Is a Key Pathogenetic Factor in Atopic Dermatitis

The histological hallmark of eczematous disorders is characterized by a marked keratinocyte pathology. Spongiosis in the epidermis is identified by impairment or loss of cohesion between KC and the influx of fluid from dermis, sometimes progressing to vesicle formation. A recent study by Trautmann et al. delineated activated skin-infiltrating T cell-induced epidermal keratinocyte apoptosis as a key pathogenic event in eczematous disorders (Trautmann et al. 2000). IFN-γ released from activated T cells upregulates Fas (CD95) on keratinocytes, which renders them susceptible to apoptosis. When the Fas number on keratinocytes reaches a threshold of approximately 40,000 Fas molecules per keratinocyte, the cells become susceptible to apoptosis. Keratinocytes exhibit a relatively low threshold for IFN-γ-induced Fas expression (0.1–1 ng/ml). This requirement is substantially achieved by low IFN-γ secreting T cells that also produce high amounts of IL-5 and IL-13 and thereby contribute to eosinophilia and IgE production (Trautmann et al. 2000). The lethal hit is delivered to keratinocytes by Fas ligand expressed on the surface of T cells that invade the epidermis and soluble Fas ligand released from T cells. In these studies, the involvement of cytokines other than IFN-γ was eliminated by experiments with different cytokines and anticytokine neutralizing antibodies. In addition, apoptosis pathways other than the Fas-pathway were ruled out by blocking T cell-induced keratinocyte apoptosis with caspase inhibitors and soluble Fas-Fc protein. Keratinocyte apoptosis was demonstrated in situ in lesional eczematous skin and patch test lesions of both AD and allergic contact dermatitis. Exposure of normal human skin and cultured skin equivalents to activated T cells demonstrated that keratinocyte apoptosis caused by skin-infiltrating T cells represents a key event in the pathogenesis of eczematous dermatitis (Trautmann et al. 2000). These studies demonstrate that both CD4$^+$ and CD8$^+$ T cells may play a role in keratinocyte injury according to their activation status. A direct contact of T cell to keratinocyte is not always required and soluble Fas ligand released from activated T cells can also induce keratinocyte apoptosis if keratinocytes are susceptible to apoptosis. IFN-γ appears to be a decisive cytokine to render keratinocytes susceptible to apoptosis.

Spongiosis is a characteristic histopathological appearance in eczematous dermatitis. It is characterized by condensation of the cells, widening of the intercellular space, and stretching of remaining intercellular contacts, resulting in a sponge-like appearance of the tissue. Homophilic interactions of the cadherin superfamily of molecules provides interkeratinocyte adhesiveness in the epidermis. Interestingly, during the early phase of keratinocyte apoptosis one of these cadherin superfamily molecules, E-cadherin, is rapidly cleaved whereas desmosomal cadherins (desmocollin and desmoglein) remain intact. Accordingly, loss of E-cadherin contacts and sustained desmosomal cadherin contacts between keratinocytes result in spongioform morphology in the epidermis (Trautmann et al. 2001 a–c). In addition, it has been demonstrated that targeting apoptosis of epidermal keratinocytes may open a new future for drug development in the treatment of asthma and AD. Current treatments such as corticosteroides, cylosporine A, rapamycine, and FK506 mainly inhibit activation of T cells and T cell-induced keratinocyte apoptosis (Trautmann et al. 2001 a). Similar apoptotic mechanisms leading to bronchial epithelial cell death were also demonstrated in asthma (Trautmann et al. 2002).

Acknowledgements. The authors' laboratory is supported by the Swiss National Foundation Grant No:32.65661.01.

Abbreviations

CLA   Cutaneous lymphocyte-associated antigen
AD   Atopic dermatitis
SEB   Staphylococcal enterotoxin B
PSGL   P-selectin glycoprotein ligand
FucT-VII α-(1,3)-Fucosyltransferase
PLA   Phospholipase A$_2$

References

Adams DH, Hubscher SG, Fisher NC, Williams A, Robinson M (1996) Expression of E-selectin and E-selectin ligands in human liver inflammation. Hepatology 24:533–538

Akdis CA, Akdis M, Simon D, Dibbert B, Weber M, Gratzl S, Kreyden O, Disch R, Wüthrich B, Blaser K, Simon H-U (1999a) T cells and T cell-derived cytokines as pathogenic factors in the nonallergic form of atopic dermatitis. J Invest Dermatol 113:628–634

Akdis M, Simon H-U, Weigl L, Kreyden O, Blaser K, Akdis CA (1999b) Skin homing (cutaneous lymphocyte-associated antigen-positive) CD8+ T cells respond to superantigen and contribute to eosinophilia and IgE production in atopic dermatitis. J Immunol 163:466–475

Akdis CA, Akdis M, Trautmann A, Blaser K (2000a) Immune regulation in atopic dermatitis. Curr Opin Immunol 12:641–646

Akdis M, Klunker S, Schliz M, Blaser K, Akdis CA (2000b) Expression of cutaneous lymphocyte-associated antigen on human CD4+ and CD8+ Th2 cells. Eur J Immunol 30:3533–3541

Akdis M, Akdis CA, Weigl L, Disch R, Blaser K (1997) Skin-homing, CLA+ memory T cells are activated in atopic dermatitis and regulate IgE by an IL-13-dominated cytokine pattern. IgG4 counter-regulation by CLA− memory T cells. J Immunol 159:4611–4619

Austrup F, Vestweber D, Borges EL, Lohning M, Bräuer R, Herz U, Renz H, Hallmann R, Scheffold A, Radbruch A, Hamann A (1997) P- and E-selectin mediate recruitment of T helper 1 but not of T helper 2 cells into inflamed tissues. Nature 385:81–84

Blander JM, Visintin I, Janeway Jr. CA, Medzhitov R (1999) $\alpha(1,3)$-Fucosyltrasferase VII and $\alpha(2,3)$-sialyltransferase IV are upregulated in activated CD4 T cells and maintained after their differentiation into Th1 and migration into inflammatory sites. J Immunol 163:3746–3752

Bos JD, Kapsenberg ML (1993) The skin immune system: progress in cutaneous biology. Immunol Today 14:75–79

Conlon K, Osborne J, Morimoto C, Ortaldo JR, Young HA (1995) Comparison of lymphokine secretion and mRNA expression in the CD45RA+ and CD45RO+ subsets of human peripheral blood CD4+ and CD8+ lymphocytes. Eur J Immunol 25:644–648

Fuhlbrigge RC, Kieffer JD, Armerding D, Kupper TS (1997) Cutaneous lymphocyte antigen is a specialized form of PSGL-1 expressed on skin homing T cells. Nature 389:978–981

Grewe J, Gyufko K, Schöpf K, Krutmann J (1994) Lesional expression of interferon-g in atopic eczema. Lancet 343:25–26

Hamid Q, Boguniewicz M, Leung DYM (1994) Differential in situ cytokine gene expression in acute versus chronic atopic dermatitis. J Clin Invest 94:870–876

Hanifin JM, Rajka G (1980) Diagnostic features of atopic dermatitis. Acta Derm Venerol 92:44–47

Heald PW, Yan SL, Edelson RL, Tigelaar R, Picker LJ (1993) Skin-selective lymphocyte homing mechanisms in the pathogenesis of leukemic cutaneous T-cell lymphoma. J Invest Dermatol 101:222–226

Herz U, Schnoy N, Borelli S, Weigl L, Käsbohrer U, Daser A, Wahn U, Köttgen R, Renz H (1998) A hu-SCID mouse model for allergic immune responses: Bacterial superantigen enhances skin inflammation and suppresses IgE production. J Invest Dermatol 110:224–231

Jones SM, Dixey J, Hall ND, McHugh NJ (1997) Expression of cutaneous lymphocyte antigen and its counter-receptor E-selectin in the skin and joints of patients with psoriatic arthritis. B J Rheum 36:748–757

Kemeny DM, Noble A, Holmes BJ, Diaz Sanches D (1994) Immune regulation: a new role for CD8⁺ T cell. Immunol Today 15:107–110

Leung DYM, Bhan AK, Schneeberger EE, Geha RS (1983) Characterization of the mononuclear cell infiltrate in atopic dermatitis using monoclonal antibodies. J Allergy Clin Immunol 71:47–55

Leung DYM, Cotran RS, Pober JS (1991) Expression of an endothelial leukocyte adhesion molecule (ELAM-1) in elicited late phase allergic skin reactions. J Clin Invest 87:1805–1810

Leung DYM, Gately M, Trumble A, Ferguson-Darnell B, Schlievert PM, Picker LJ (1995) Bacterial superantigens induce T cell expression of the skin-selective homing receptor, the cutaneous lymphocyte-associated antigen, via stimulation of interleukin 12 production. J Exp Med 181:747–753

Leyden JE, Marpies RR, Kligman AM (1974) *Staphylococcus aureus* in the lesions of atopic dermatitis. B J Derm 90:525–530

Lim Y-C, Henault L, Wagers AJ, Kansas GS, Luscinskas FW, Lichtman AH (1999) Expression of functional selectin ligands on Th cells is differentially regulated by IL-12 and IL-4. J Immunol 162:3193–3201

Mosmann TR and Sad S (1996) The expanding universe of T-cell subsets: Th1, Th2 and more. Immunol Today 17:142–146

Picker LJ, Kishimoto TK, Smith CW, Warnock RA, Butcher EC (1991) ELAM-1 is an adhesion molecule for skin homing T cells. Nature 349:796–799

Picker LJ, Michie SA, Rott LS, Butcher EC (1990) A Unique phenotype of skin associated lymphocytes in humans: preferential expression of the HECA-452 epitope by benign and malignant T-cells at cutaneous sites. Am J Pathol 136:1053–1061

Picker LJ, Treer JR, Ferguson-Darnell B, Collins PA, Bergstresser PR, Terstappen LWMM (1993) Control of lymphocyte recirculation in man. III. Differential regulation of the cutaneous lymphocyte-associated antigen, a tissue selective homing receptor for skin-homing T cells. J Immunol 150:1122–1136

Rook AH, Kang K, Kubin M, Cassin M, Trinchieri G, Lessin SR, Cooper KD (1994) Interleukin 12 mRNA and protein production by epidermal Langerhans cells (abstract). Clin Res 42:231

Rossiter H, Mudde GC, van Reijsen F, Kalthoff F, Bruijnzeel-Komen CAFM, Picker LJ, Kupper TS (1994) Disease-related T cells from atopic skin express cutaneous lymphocyte antigen and sialyl Lewis X determinants, and bind to both E-selectin and P-selectin. Eur J Immunol 24:205–210

Sallusto F, Mackay CR, Lanzavecchia A (1997) Selective expression of the eotaxin receptor CCR3 by human T helper 2 cells. Science 277:2005–2007

Santamaria Babi LF, Moser R, Perez Soler MT, Picker LJ, Blaser K, Hauser C (1995a) The migration of skin-homing T cells across cytokine-activated human endothelial cell layers involves interaction of the cutaneous lymphocyte-associated antigen (CLA), the very late antigen-4 (VLA-4) and the lymphocyte function-associated antigen-1 (LFA-1). J Immunol 154:1543–1550

Santamaria Babi LF, Picker LJ, Perez Soler MT, Drzimalla K, Flohr P, Blaser K, Hauser C (1995b) Circulating allergen-reactive T cells from patients with atopic dermatitis and allergic contact dermatitis express the skin-selective homing receptor, the cutaneous lymphocyte-associated antigen. J Exp Med 181:1935–1940

Sasaki K, Kurata K, Funayama K, Nagata M, Watanabe E, Ohta S, Hanai N, Nishi T (1994) Expression cloning of a novel α1,3-fucosyltransferase that is involved in biosynthesis of the sialyl lewis X carbohydrate determinants in leukocytes. J Biol Chem 269:14730–14737

Thepen T, Langeveld-Wildschut EG, Bihari IC, van Vichen DF, Van Reijsen FC, Mudde GC, Bruijnzeel-Koomen CAFM (1996) Biphasic response against aeroallergen in atopic dermatitis showing a switch from an initial Th2 response to a Th1 response in situ: An immunohistochemical study. J Allergy Clin Immunol 97:828–837

Trautmann A, Akdis M, Schmid-Grendelmeier P, Dissch R, Bröcker E-B, Blaser K, Akdis CA (2001a) Targeting keratinocyte apoptosis in the treatment of atopic dermatitis and allergic contact dermatitis. J. Allergy Clin Immunol. 108:839–846

Trautmann A, Akdis M, Brocker EB, Blaser K, Akdis CA (2001b) New insights into the role of T cells in atopic dermatitis and allergic contact dermatitis. Trends Immunol 22:530–2

Trautmann A, Akdis M, Kleeman D, Altznauer F, Simon H-U, Graeve T, Noll M, Blaser K, Akdis CA (2000c) T cell-mediated Fas-induced keratinocyte apoptosis plays a key pathogenetic role in eczematous dermatitis. J Clin Invest 106:25–35

Trautmann A, Altznauer F, Akdis M, Simon H-U, Disch R, Bröcker E-B, Blaser K, Akdis CA (2001) The differential fate of cadherins during T cell-induced keratinocyte apoptosi leads to spongiosis in ectematous dermatitis. J Invest Derm 117:927–934

Trautmann A, Schmid-Grendelmeier P, Krüger K, Crameri R, Akdis M, Akkaya A, Bröcker E-B, Blaser K, Akdis AC (2002) T cells and eosinophils cooperate in the induction of bronchial epithelial apoptosis in asthma. J Allergy Clin Immunol 109:329–337

Wagers AJ, Waters CM, Stoolman LM, Kansas GS (1998) Interleukin 12 and interleukin 4 control T cell adhesion to endothelial selectins through opposite effects on α1,3-fucosyltransferase VII gene expression. J Exp Med 188:2225–2231

Wüthrich B (1978) Serum IgE in atopic dermatitis. Clin Allergy 8:241–248

3 Cutaneous Lymphocyte Localization in the Pathogenesis of Psoriasis

M. P. Schön

3.1 Psoriasis – A Complex Inflammatory Skin Disorder with a T Cell-Based Immunopathogenesis

Psoriasis is one of the most common chronic inflammatory skin diseases, affecting 1%–3% of the Caucasian population worldwide (Barker 1994; Christophers 1996). This complex disease is characterized by pathological changes in a variety of different cell types. These include epidermal keratinocyte hyperproliferation and altered differentiation as indicated by parakeratosis (nuclei in the stratum corneum), aberrant expression of the hyperproliferation-associated keratin pair 6/16 (Stoler et al. 1988; Weiss et al. 1984), involucrin and filaggrin (Bernard et al. 1986; Ishida-Yamamoto and Iizuka 1995), and integrin adhesion molecules (Hertle et al. 1992; Kellner et al. 1992). In addition, de novo expression of major histocompatibility complex (MHC) class II and intercellular adhesion molecule-1 (ICAM-1, CD54) by keratinocytes is observed (Barker et al. 1990; Nickoloff et al. 1990; Veale et al. 1995), i.e., molecules involved in interactions with immigrating T lymphocytes. Endothelial cells also are hyperproliferative, resulting in angiogenesis and dilation of dermal blood vessels, and express increased levels of ICAM-1, E-selectin (CD62E) and vascular cell adhesion molecule-1 (VCAM-1, CD106), as well as MHC class II, indicating activation (Detmar et al. 1994; Goodfield et al. 1994; Bjerke et al. 1988; Das et al. 1994). Finally, a mixed leukocytic infiltrate is seen composed of activated T lymphocytes (Ramirez-Bosca et al. 1988; Schlaak et al. 1994), neutrophils within the dermis and forming the telltale Munro's microabscesses within the epidermis (Pinkus and Mehregan 1966), and an increased number of dermal mast cells and dendritic cells (Rothe et al. 1990; van de Kerkhof et al. 1995). A complex network of cytokines, chemokines, and other mediators is thought to mediate the psoriatic tissue alterations (Nickoloff 1991; Schön and Ruzicka 2001). Since its complex pathogenesis requires careful orchestration of sequential and highly specific leukocyte functions, psoriasis may serve as a model disease for studying the intertwined interactions of immigrating immune cells with resident epithelial and mesenchymal cells.

While many factors leading to the generation of psoriatic lesions still remain obscure, compelling circumstantial evidence is accumulating indicating a primary T lymphocyte-based immunopathogenesis (Barker 1994; Christophers 1996). This evidence is based upon the response of psoriasis to treatment with lymphocyte-specific compounds, such as cyclosporin A (Mueller and Herrmann 1979), the toxin $DAB_{389}IL-2$ (Gottlieb et al. 1995), compounds targeting CD2 (Aruffo and Hollenbaugh 2001; Ellis and Krueger 2001; Krueger 2002), or, in some cases, anti-CD4 monoclonal antibodies (Nicolas et al. 1991, 1992; Prinz et al. 1991). In addition, there is a possible linkage of a psoriasis susceptibility gene with a gene involved in interleukin-2 (IL-2) regulation (Tomfohrde et al. 1994), and psoriasis does not recur after transplantation of bone marrow from healthy donors (Eedy et al. 1990). Furthermore, the association of psoriasis with certain MHC alleles, such as -B13, -B17, -Bw57, and -Cw6

(Nickoloff 1999) suggests a pathogenic role of T cells. While observations that eruption of psoriatic skin lesions coincides with epidermal infiltration and activation of T cells, that resolution of the lesions is preceded by reduction or disappearance of those T cells (Valdimarsson et al. 1997), and that lesional psoriatic T cells may alter keratinocyte differentiation and antigen expression (Skov et al. 1997; Strange et al. 1993) do not formally prove a primary role of T cells, they at least support this concept. Some investigators have reported a restricted T cell receptor (TCR) variable gene usage of T lymphocytes within psoriatic lesions (Boehncke et al. 1994; Bour et al. 1999; Lewis et al. 1993; Menssen et al. 1995), a finding that strongly suggests an antigen-specific response of T lymphocytes. While the pathogenetic relevance of this oligoclonal T cell expansion is not entirely clear (Vekony et al. 1997), it is possible that the failure to demonstrate oligoclonality in some cases of psoriasis is due, at least in part, to colonization of psoriatic lesions by superantigen-producing bacteria (Travers et al. 1999). Given the well-established role of bacterial superantigens in the pathogenesis of psoriasis (Leung et al. 1995 a, b; Travers et al. 1999) and recently identified sequence similarities between streptococcal M-peptides and human epidermal keratins, such as keratin 17, it is indeed possible that keratinocyte structural proteins function as autoantigens in the psoriatic disease process (Bonnekoh et al. 2001; Gudmundsdottir et al. 1999; Valdimarsson et al. 1997).

Although there is no naturally occurring animal disease mirroring psoriasis, additional support for a primary role of T lymphocytes in the pathogenesis of psoriasis comes from animal studies. Indeed, in studies using a xenotransplantation model of psoriasis, injection of T lymphocytes from psoriasis patients into unaffected skin transplanted from the same patients onto severe combined immunodeficiency (*scid/scid*) mice resulted in the generation of psoriatic skin lesions (Wrone-Smith and Nickoloff 1996). In addition, bacterial superantigens apparently stimulate this pathogenic function of T cells (Boehncke et al. 1996). In another rodent model of psoriasis, transfer of minor histocompatibility-mismatched CD4$^+$/CD45RB^{hi} T lymphocytes into *scid/scid* mice resulted in the generation of psoriasiform skin lesions in the absence of a primary epithelial abnormality (Schön et al. 1997). Bacterial superantigens again enhance the disease severity in this model (Hong et al. 1999), and coinjection of previously activated $CD4^+/CD45RB^{lo}$ T cells or transfer of unfractionated T lymphocytes suppresses the skin lesions, indicating that psoriasiform lesions in this model are based upon T cell-mediated immune dysregulation (Schön et al. 1997). Finally, in HLA-B27-transgenic rats it has been demonstrated that the inflammatory disorder including psoriasiform skin lesions is initiated by T lymphocytes without pre-existing epithelial abnormalities (Breban et al. 1996).

3.2 Chemokines and Cytokines – Road Signs for Cutaneous Recruitment of Lymphocytes in Psoriasis

Few diseases illustrate the central role of cytokines, chemokines, and other mediators for tissue-specific leukocyte recruitment as vividly as psoriasis (Nickoloff 1988; Schön and Ruzicka 2001; schematically depicted in Fig. 1). Intracutaneous secretion and/or pathological dysregulation of a number of cytokines is thought to mediate crucial cellular interactions resulting in the tissue alterations seen in psoriasis (Nickoloff 1991). These cytokines include tumor necrosis factor-a (TNF-a) and interleukin-1 (IL-1; Kupper 1990), interferon-γ (IFN-γ; Barker et al. 1991; Gottlieb et al. 1988 b), IL-6 (Castells-Rodellas et al. 1992; Neuner et al. 1991), vascular endothelial growth factor (VEGF; Detmar et al. 1994), and transforming growth factor-a (TGF-a; Elder et al. 1989; Gottlieb et al. 1988 a; Prinz et al. 1994). These cytokines induce a number of adhesion molecules, such as ICAM-1, VCAM-1, or $\beta 1$ integrins which mediate adhesive interactions with immigrating leukocytes. Dysregulation of the immunosuppressive cytokine, IL-10, also has been implicated in the pathogenesis of psoriasis (Michel et al. 1996, 1997), a concept that is strongly supported by the therapeutic efficacy of IL-10 in some cases of psoriasis (Asadullah et al. 1998, 1999). The microenvironmental regulation of these pathogenic events, however, still remains somewhat enigmatic.

Recent evidence strongly suggests a pivotal role of chemokines for trafficking, adhesion, subtype-specific influx, and compartmentalization of leukocytes in the psoriatic disease process (Homey et al. 2000a, 2002; Schön and Ruzicka 2001; Zlotnik et al. 1999): The

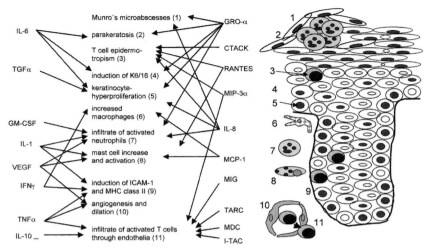

Fig. 1. Selected cytokines (*left*) and chemokines (*right*) thought to be involved in tissue localization of infiltrating leukocytes and the pathogenesis of psoriasis are schematically depicted

chemokines TARC and MDC (macrophage-derived chemokine), on the one hand, are both expressed by the cutaneous vasculature and bind to the CCR4 receptor. CCR4, on the other hand, is expressed by some circulating CLA$^+$ memory T cells, suggesting that these chemokines contribute to the preferential recruitment of skin-homing memory T cells via stimulation of integrin/ICAM-1 adhesive interactions (Campbell et al. 1999).

CTACK (cutaneous T cell-attracting chemokine, CCL27) has been identified as a ligand for the orphan chemokine receptor GPR-2 (CCR10). This chemokine is constitutively expressed by epidermal keratinocytes, but can be significantly increased by TNF-α and IL-1β, cytokines thought to be involved in the psoriatic disease process (Homey et al. 2000b). Given that CCR10 is expressed by T cells and skin-derived Langerhans cells, CTACK appears to contribute to epidermal localization of these cells (Homey et al. 2002).

Colocalization of MIP-3α (macrophage inflammatory protein-3α, CCL20) with epidermal T cells has been described in psoriatic epidermis. Functional relevance of this finding is suggested by the observation that skin-homing CLA$^+$ T lymphocytes express high levels of the MIP-3α-receptor, CCR6 (Homey et al. 2000). As compared to T cells from normal donors, psoriatic CLA$^+$ T cells respond to lower concentrations of MIP-3α. In addition, MIP-3α can be induced on keratinocytes by type I proinflammatory cytokines thought to be involved in the pathogenesis of psoriasis (Homey et al. 2000a).

The CXC-chemokine, MIG (monokine induced by γ-interferon), is expressed almost exclusively by a spatially restricted subpopulation of endothelial cells and macrophages within the papillary dermis directly underneath the hyperplastic psoriatic epidermis (Goebeler et al. 1998). As MIG is a T cell-attracting chemokine (Liao et al. 1995), it may contribute to epidermal T cell localization in psoriatic skin, possibly through TGF-β_1-induced expression of the α_E(CD103)β_7 integrin (Agace et al. 2000). It is also possible, however, that MIG is involved in localization of T cells to the papillary dermis. It has been demonstrated that MIG can be induced in macrophages and dermal microvascular endothelial cells by T cell-derived IFN-γ (Goebeler et al. 1998; Farber 1993). It is, therefore, conceivable that a microenvironmental T-cell-associated, inflammation-boosting loop contributes to the histopathological changes of psoriatic skin. This pathogenic process may be enhanced by RANTES (regulated on activation, normal T cell expressed and secreted), a C-C lymphocyte-attracting chemokine that is upregulated preferentially within psoriatic epidermis (Raychaudhuri et al. 1999).

Another loop may exist with respect to MCP-1 (monocyte chemotactic protein-1), another CC-chemokine whose expression is upregulated in psoriatic epidermis (Vestergaard et al. 1997). MCP-1 stimulates its own production by monocytes (autocrine loop), and fibroblast expression of TGF-β_1 (Yamamoto et al. 2000), a cytokine thought to contribute to epidermal T cell localization through induction of the integrin α_E(CD103)β_7 (Pauls et al. 2001). In addition, both MCP-1 and RANTES may attract mast cells to psoriatic skin (Raychaudhuri et al. 1999). Since the CXC chemokine, I-TAC (IFN-inducible T cell α chemoattractant), can be induced by IFN-γ on keratinocytes and endothelial cells (Mazanet et al. 2000), it is conceivable that I-TAC also contributes to cutaneous T cell localization in psoriasis. However, there is no direct evidence at present to support this notion.

The formation of epidermal neutrophil accumulations, termed Munro's microabscesses, is a hallmark feature of psoriasis (Christophers 1996). Psoriatic scales, the product of the excessive keratinization within the psoriatic plaques, contain high amounts of neutrophil-attracting chemokines including IL-8 and GRO-a (growth related cytokine-a; Kulke et al. 1996; Gillitzer et al. 1996). This results in a vicious circle of neutrophil attraction and further tissue damage mediated by neutrophils themselves as well as other mechanisms. In addition IL-8 downregulates epidermal expression of the IL-10 receptor, resulting in a loss of negative immunoregulatory stimuli in psoriatic lesions (Michel et al. 1996, 1997). Keratinocyte hyperproliferation in psoriatic skin also appears to be mediated, at least in part, by IL-8 and GRO-a, as their receptor, CXCR2, is overexpressed in psoriatic but not normal skin (Kulke et al. 1998).

3.3 Adhesion Molecules – Molecular Ladders for Cutaneous Lymphocyte Recruitment

3.3.1 Tethering and Rolling

Tissue-selective trafficking of T lymphocytes is mediated by complex interactions of cytokines/chemokines and various adhesion receptors (Kunkel and Butcher 2002). Adhesion molecules are crucial for the site-specific recruitment and functions of T cells in most, if not all, inflammatory skin diseases (schematically depicted in Fig. 2). While truly psoriasis-specific adhesive interactions have not been identified thus far, several pathogenic steps mediated by adhesion molecules have been studied in psoriasis and psoriasis has been used as a target disorder to treat inflammation by interfering with adhesion molecule functions.

The first steps of T cell localization to all tissues include leukocyte rolling on the vessel wall mediated primarily by selectins (Butcher and Picker 1996; Groves et al. 1991; Shimizu et al. 1991; Smith et al. 1993; Springer 1994; von Andrian et al. 1991; von Andrian and Mackay 2000; Robert and Kupper 1999). Selectins are single-chain transmembrane adhesion molecules characterized by a lectin-like domain that binds to carbohydrate ligands displayed on gly-

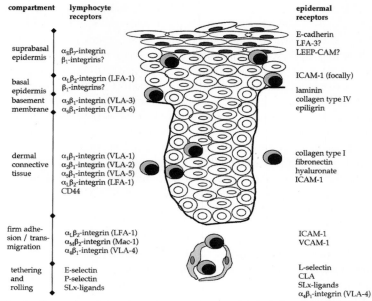

Fig. 2. Synopsis of selected adhesion molecules thought to mediate cutaneous localization in psoriasis and other inflammatory skin disorders

coprotein scaffolds (Feizi 2001; Ley 2001; Varki 1994). Activated endothelial cells, on the one hand, rapidly mobilize P-selectin (CD62P) to the cell surface (Bonfanti and Furie 1989; McEver and Beckstead 1989), while expression of E-selectin (CD62E) is transcriptionally regulated (Bevilaqua et al. 1991; Cotran and Gimbrone 1986). The pivotal role of P- and E-selectin for leukocyte rolling has been confirmed by a large variety of experimental approaches interfering with adhesive interactions of selectins and their carbohydrate ligands (Carlos and Harlan 1994; Todderud et al. 1997). However, in a recent clinical trial, a function-blocking antibody directed against E-selectin did not alleviate psoriasis (Bhushan et al. 2002). This observation suggested that some selectin-mediated functions may be redundant and that interfering with a single selectin alone is not sufficient to interrupt the inflammatory chain in psoriasis and, presumably, other inflammatory diseases. This notion is supported by efomycines, recently discovered specific, small-molecule inhibitors of both E- and P-selectin functions. Blocking both E- and P-selectin by efomycine M significantly inhibited rolling of T lymphocytes on cutaneous microvessels and markedly alleviated chronic psoriasiform inflammatory skin conditions in a T cell-mediated murine model of psoriasis as well as in human psoriatic skin transplanted onto *scid/ scid* mice (Schön et al. 2002). Another way to inhibit selectin-mediated adhesive interactions for treating psoriasis and, possibly, other inflammatory conditions appears to be specific interference with proteasomes, thereby inhibiting NF-κB-mediated transactivation of proinflammatory genes, expression of selectin ligands, and, consecutively, selectin-mediated binding (Zollner et al. 2002).

On the other hand, T cells express L-selectin (CD62L) that binds to endothelial cell selectin ligands. The topographic distribution of L-selectin on the tips of microvilli of rolling leukocytes appears to be important for contact formation with endothelial ligands (Fors et al. 2001). It is thought that shedding of L-selectin, a proteolytic process mediated by metalloproteases (Condon et al. 2001; Zhao et al. 2001), also plays a role for proper lymphocyte rolling (Hafezi-Moghadam and Ley 1999). Inhibition of L-selectin shedding results in increased LFA-1/ICAM-1-mediated firm adhesion and transmigration of lymphocytes (Hafezi-Moghadam et al. 2001), suggesting a regulatory function of the shedding process.

T cells express E-selectin ligands which are transmembrane glycoproteins bearing the Sialyl-Lewis[X] moiety (sLe[X]; Varki 1994). T lymphocytes localizing to the skin express the sLe[X]-bearing CLA (cutaneous lymphocyte-associated antigen), a specialized form of PSGL-1 (P-selectin glycoprotein ligand-1, CD162; Fuhlbrigge et al. 1997), which is thought to be involved in tissue-specific localization of cutaneous T cells (Picker et al. 1990, 1991). CLA-bearing T lymphocytes appear to extravasate preferentially through the endothelium of the superficial dermal plexus (in the original report demonstrated in normal and psoriatic skin; Kunstfeld et al. 1997) suggesting topographic specialization of microvascular endothelial cells within the skin.

In addition to selectin-mediated rolling, VLA-4 (very late activation antigen-4, CD49d/CD29), a heterodimeric adhesion receptor of the integrin family that binds to the immunoglobulin-superfamily adhesion molecules, including VCAM-1 (vascular cell adhesion molecule-1) and MAdCAM-1 (mucosal addressin cell adhesion molecule-1), has also been found to mediate rolling of certain leukocyte subsets (Berlin et al. 1995; Reinhardt et al. 1997; Singbartl et al. 2001). This involvement in leukocyte rolling is exerted in addition to its known function for firm adhesion (see Sect. 3.3.2) and is due, at least in part, to the topographic presentation of VLA-4 on microvilli of rolling cells, thus enabling the first contact to endothelial-bound counter-receptors (Berlin et al. 1995). VLA-4 affinity is thought to be rapidly upregulated upon T cell stimulation via signaling through the $p56^{lck}$ Src kinase pathway (Feigelson et al. 2001), a process that may be important for the transition from rolling to firm adhesion. While some aspects of the interplay of VLA-4- and selectin-mediated adhesive interactions involved in leukocyte rolling still remain to be unraveled, it appears that their relative contributions are influenced by tissue- and (micro)environment-specific factors, and that there is some redundancy in their functions. In cutaneous inflammation in rats, all three receptors, E-selectin, P-selectin, and VLA-4, were required for rolling of memory T lymphocytes (Issekutz and Issekutz 2002). Their relative roles in human psoriasis has not been determined yet.

3.3.2 Firm Adhesion to the Endothelial Lining and Extravasation

Stimulatory effects exerted by a growing number of cytokines, chemokines and other mediators initiate the subsequent adhesive steps of cutaneous lymphocyte localization (Homey et al. 2002; Schön and Ruzicka 2001; Zlotnik et al. 1999). After transient, selectin- and VLA-4-mediated rolling, leukocytes firmly attach to the endothelium through adhesion of β_2-integrins, including LFA-1 (CD11a/CD18, $a_L\beta_2$) or Mac-1 (CD11b/CD18, $a_M\beta_2$), to immunoglobulin superfamily members, such as ICAM-1 (CD54; Dustin et al. 1986; Griffiths et al. 1989; schematically depicted in Fig. 2). This mechanism appears to be of prime importance in various inflammatory skin conditions (Grabbe et al. 2002). In addition, β_1-integrins and their ligands, such as the $a_4\beta_1$/VCAM-1 pair, are involved in leukocyte-endothelial cell binding (Groves et al. 1993). Proinflammatory cytokines, including IFN-γ, TNF-α, and IL-1 can increase T cell localization to inflammatory sites through induction of ICAM-1 and VCAM-1, (Barker et al. 1990; Griffiths et al. 1989; Groves et al. 1993; Petzelbauer et al. 1994). It is possible that additional mechanisms contribute to firm adhesion and endothelial transmigration of lymphocytes, similar to a novel mechanism proposed for the $\alpha_E\beta_7$ integrin within the intestinal lamina propria (Strauch et al. 2001), but such mechanisms in psoriatic skin remain to be unraveled.

3.3.3 Dermal Lymphocyte Localization

Once extravasated at cutaneous sites, lymphocytes utilize β_1-integrins to bind to and transmigrate through the dermal extracellular matrix (ECM; schematically depicted in Fig. 2). These ECM-receptors include the $\alpha_1\beta_1$, $\alpha_2\beta_1$, and $\alpha_5\beta_1$ integrins which bind to various extracellular matrix components such as collagen type I, fibronectin, chondroitin sulfate, laminin, or hyaluronans (Hemler 1990; Hynes 1992; Konter et al. 1989). In most inflammatory skin disorders, including psoriasis, dermal lymphocytes greatly outnumber epidermal lymphocytes (Hwang 2001), suggesting that only a minority of infiltrating T cells has acquired the molecular armory to migrate into the cutaneous epithelium. Given that composition, fibril diameter, as well as three-dimensional arrangement of ECM molecules, show site-specific variation (e.g., papillary dermis, reticular dermis, perivascular or periadnexal areas may provide specific microenvironments) and may be altered profoundly in inflammatory states (Berthod et al. 2001); it appears likely that tissue-specific leukocyte localization and the distribution pattern characteristic for inflammatory skin disorders, including psoriasis, is modulated, at least in part, by such factors. However, there is no direct evidence at this time to corroborate these hypotheses.

In addition to β_1-integrins, CD44, a hyaluronate receptor (Camp et al. 1993), and LFA-1, which may interact with an "adhesive path" formed by interstitial ICAM-1 (Nickoloff 1988), appear to facilitate dermal localization of T cells. Binding to components of the epidermal basement membrane, such as collagen type IV and laminin, again appears to be mediated by β_1-integrins, including $\alpha_1\beta_1$, $\alpha_2\beta_1$, $\alpha_3\beta_1$, and $\alpha_6\beta_1$ (Hynes 1992). This adhesive interaction may be enhanced by keratinocyte-derived mediators, such as IL-7 (Wagner et al. 1999). In addition, T cells may utilize the $\alpha_3\beta_1$ integrin to bind to epiligrin within the basement membrane (Wayner et al. 1993).

3.3.4 Epidermal T Cell Localization

In contrast to endothelial and dermal localization, we know relatively little about epidermal localization of T cells (schematically depicted in Fig. 2). The epidermis of the skin is a multilayered, stratified, and polarized epithelium, whose different layers show distinct stages of differentiation and surface antigen expression. Another level of complexity is added by profound changes of the epidermal differentiation pattern under inflammatory conditions impacting on the ability of lymphocytes to localize to the epidermis. As a consequence, epidermal localization of lymphocytes appears to be a rather complex process in itself, and certain lymphocyte subsets may specifically localize only to particular layers of the epidermis. Such a "subcompartmentalization" of epidermal lymphocytes is suggested by the observation that most epidermal T cells reside within the basal layer or adjacent to the dermo-epidermal junction (the basement membrane zone), while relatively few migrate to suprabasal (micro)-compartments in most epidermotropic inflammatory conditions or cutaneous T cell lymphomas. This is reflected by the expression pattern and spatial distribution of adhesion molecules involved in interactions with lymphocytes. However, it is also possible that some lymphocytes are transported passively to upper epidermal layers due to the high turnover and, consecutively, increased upward migration of keratinocytes in the psoriatic epidermis.

While many extracellular ligands for β_1-integrins are not expressed beyond the epidermal basement membrane (Konter et al. 1989), β_1 integrins have been implicated in T cell epidermotropism based upon expression by intraepidermal T lymphocytes (Sterry et al. 1992). An example is the $\alpha_1\beta_1$ integrin, whose expression by lymphocytes appears to be associated with epidermotropic forms of cutaneous T cell lymphoma (CTCL), but whose functional contribution to the process of epidermotropism remains unclear (Bank et al. 1999). In psoriasis, a significant upregulation of integrins $\alpha_1\beta_1$, $\alpha_2\beta_1$, and $\alpha_3\beta_1$ on intraepidermal T lymphocytes has been observed by two-color FACS analysis (M. P. Schön et al., unpublished data). The functional relevance of this finding, however, remains to be unraveled.

Induced by proinflammatory cytokines, there is epidermal de novo expression of ICAM-1 in inflammatory skin disorders, such as psoriasis (Dustin et al. 1988; Griffiths et al. 1989). Indeed, in vitro studies suggested that ICAM-1/LFA-1 interactions are involved in binding of activated T cells to inflamed epidermis (Kashihara-Sawami and Norris 1992). Given that ICAM-1 is induced primarily in basal keratinocytes upon inflammatory stimuli, it appears to be involved in the initial steps of epidermal T cell localization just beyond the epidermal basement membrane. However, this may not be the only mechanism, since constitutive epidermal expression of ICAM-1 in transgenic mice did not result in spontaneous epidermal T cell infiltration (Williams and Kupper 1994), and expression and spatial distribution of ICAM-1 and LFA-1 do not correlate in many cases of epidermotropic lymphocyte infiltration (Griffiths et al. 1989; Olivry et al. 1995). In addition, the interaction of LFA-1 expressed by lymphocytes with ICAM-3, which is constitutively expressed by epidermal keratinocytes, may be involved in epidermal T cell localization (Griffiths et al. 1995).

The recently identified glycoprotein LEEP-CAM (lymphocyte endothelial-epithelial-cell adhesion molecule), whose ligand on T lymphocytes has not been identified yet, may also be involved in epidermal T cell localization (Shieh et al. 1999). The LEEP-CAM molecule, a 90-kDa to 115-kDa cell surface glycoprotein, mediates T cell adhesion to epithelial cells in static cell-to-cell adhesion assays in vitro (Shieh et al. 1999). It is expressed constitutively within the suprabasal epidermal layers in both normal and psoriatic skin, but is not expressed on T cells. Both its expression pattern and its adhesive functions in vitro make LEEP-CAM an interesting candidate molecule for mediating epidermal T cell localization. Since LEEP-CAM is expressed exclusively within suprabasal epidermal layers both in normal and psoriatic skin (Shieh et al. 1999), it may be the first adhesion receptor that preferentially mediates suprabasal localization of T cells.

Another player contributing to epidermal localization of certain T cell subsets is the $\alpha_E(CD103)\beta_7$ integrin, which is expressed by most intestinal intraepithelial T lymphocytes (Kilshaw and Baker 1988; Parker et al. 1992), and is thought to contribute to localization of diffusely distributed T cell subsets to the intestinal epithelium through binding to E-cadherin (Cepek et al. 1994; Karecla et al. 1995). Indeed, when integrin $\alpha_E(CD103)$-deficient mice were studied, they exhibited a reduced number of mucosal intraepithelial T cells (Schön et al. 1999). However, there is growing evidence that $\alpha_E(CD103)\beta_7$ functions are not restricted to T cells within the intestinal mucosa, but also occur in intraepithelial T lymphocytes in other tissues, whose epithelia express E-cadherin, the ligand for $\alpha_E(CD103)\beta_7$ (Agace et al. 2000). Putative alternative ligand(s) for $\alpha_E\beta_7$ on epithelial (Brown et al. 1999) and endothelial cells (Strauch et al. 2001) have been proposed but not positively identified yet. In the skin, expression of $\alpha_E(CD103)\beta_7$ has been demonstrated on epidermal T lymphocytes in several inflammatory disorders (de Vries et al. 1997; Walton et al. 1997) and cutaneous T cell lymphomas (Schechner et al. 1999; Simonitsch et al. 1994). In a recent study, preferential expression of $\alpha_E(CD103)\beta_7$ was demonstrated on epidermal CD8$^+$ T cells within psoriatic lesions (Pauls et al. 2001), and there is a limited association of $\alpha_E\beta_7$ expression with expression of the chemokine receptors CXCR3 and CCR4 (Rottman et al. 2001). T cells expressing $\alpha_E\beta_7$ were located within both the basal and suprabasal epidermal layers, consistent with the expression of its ligand, E-cadherin, which is distributed throughout the viable layers of the epidermis (Pauls et al. 2001). Expression of $\alpha_E(CD103)\beta_7$ was detected on very few dermal T cells in psoriatic lesions as well as in the peripheral blood. Thus, $\alpha_E(CD103)\beta_7$ appears to be induced on CD8$^+$ T cells in situ upon entering the epidermis, consistent with the focal overexpression of TGF-β_1 directly underneath the epidermis. In further support of an involvement in T cell epidermotropism, $\alpha_E(CD103)\beta_7$ could be specifically upregulated by TGF-β_1 on CD8$^+$ T cells where it mediated adhesion to psoriatic epidermis, as well as to cultured keratinocytes (Pauls et al. 2001).

References

Agace WW, Higgins JM, Sadasivan B, Brenner MB, Parker CM (2000) T-lymphocyte-epithelial-cell interactions: integrin alpha(E)(CD103)beta(7), LEEP-CAM and chemokines. Curr Opin Cell Biol 12:563–568

Aruffo A, Hollenbaugh D (2001) Therapeutic intervention with inhibitors of co-stimulatory pathways in autoimmune disease. Curr Opin Immunol 13:683–686

Asadullah K, Sterry W, Stephanek K, Jasulaitis D, Leupold M, Audring H, Volk HD, Docke WD (1998) IL-10 is a key cytokine in psoriasis. Proof of principle by IL-10 therapy: a new therapeutic approach. J Clin Invest 101:783–794

Asadullah K, Docke WD, Ebeling M, Friedrich M, Belbe G, Audring H, Volk HD, Sterry W (1999) Interleukin 10 treatment of psoriasis: clinical results of a phase 2 trial. Arch Dermatol 135:187–192

Bank I, Rapman E, Shapiro R, Schiby G, Goldberg I, Barzilai A, Trau H, Gur H (1999) The epidermotropic mycosis fungoides associated alpha1beta1 integrin (VLA-1, CD49a/CD29) is primarily a collagen IV receptor on malignant T cells. J Cutan Pathol 26:65–71

Barker JNWN (1994) The immunopathology of psoriasis. Bailliere's Clin Rheumatol 8:429–437

Barker JNWN, Karabin GD, Stoof TJ, Sarma VJ, Dixit VM, Nickoloff BJ (1991) Detection of interferon-gamma mRNA in psoriatic epidermis by polymerase chain reaction. J Dermatol Sci 2:106–111

Barker JNWN, Sarma V, Mitra RS, Dixit VM, Nickoloff BJ (1990) Marked synergism between tumor necrosis factor-alpha and interferon-gamma in regulation of keratinocyte-derived adhesion molecules and chemotactic factors. J Clin Invest 85:605–608

Berlin C, Bargatze RF, Campbell JJ, von Andrian UH, Szabo MC, Hasslen SR, Nelson RD, Berg EL, Erlandsen SL, Butcher EC (1995) alpha 4 integrins mediate lymphocyte attachment and rolling under physiologic flow. Cell 80:413–22

Bernard BA, Reano A, Darmon YM, Thivolet J (1986) Precocious appearance of involucrin and epidermal transglutaminase during differentiation of psoriatic skin. Br J Dermatol 114:279–283

Berthod F, Germain L, Li H, Xu W, Damour O, Auger FA (2001) Collagen fibril network and elastic system remodeling in a reconstructed skin transplanted on nude mice. Matrix Biol 20:463–473

Bevilaqua PM, Edelson RL, Gasparro FP (1991) High performance liquid chromatography analysis of 8-methoxypsoralen monoadducts and cross-links in lymphocytes and keratinocytes. J Invest Dermatol 97:151–155

Bhushan M, Bleiker TO, Ballsdon AE, Allen MH, Sopwith M, Robinson MK, Clarke C, Weller RP, Graham-Brown RA, Keefe M, Barker JN, Griffiths CE (2002) Anti-E-selectin is ineffective in the treatment of psoriasis: a randomized trial. Br J Dermatol 146:824–831

Bjerke JR, Livden JK, Matre R (1988) Fc gamma-receptors and HLA-DR antigens on endothelial cells in psoriatic skin lesions. Acta Derm Venereol 68:306–11

Boehncke WH, Dressel D, Zollner TM, Kaufmann R (1996) Pulling the trigger on psoriasis (letter). Nature 379:777

Boehncke WH, Kuenzlen C, Zollner T, Mielke V, Kaufmann R (1994) Predominant usage of distinct T-cell receptor V beta regions by epidermotropic T cells in psoriasis. Exp Dermatol 3:161–163

Bonfanti R, Furie BC (1989) PADGEM (GMP140) is a component of Waibel-Palade bodies of human endothelial cells. Blood 73:1109–1112

Bonnekoh B, Böckelmann R, Ambach A, Gollnick H (2001) Dithranol and dimethylfumarate suppress the interferon-gamma-induced up-regulation of cytokeratin 17 as a putative psoriasis autoantigen in vitro. Skin Pharmacol Appl Skin Physiol 14:217–225

Bour H, Puisieux I, Kouritsky P, Favrot M, Musette P, Nicolas JF (1999) T-cell repertoire analysis in chronic plaque psoriasis suggests an antigenspecific immune response. Hum Immunol 60:665–676

Breban M, Fernandez-Sueiro JL, Richardson JA, Hadavand RR, Maika SD, Hammer RE, Taurog JD (1996) T cells, but not thymic exposure to HLA-B27, are required for the inflammatory disease of HLA-B27 transgenic rats. J Immunol 156:794–803

Brown DW, Furness J, Speight PM, Thomas GJ, Li J, Thornhill MH, Farthing PM (1999) Mechanisms of binding of cutaneous lymphocyte-associated antigen-positive and alphaEbeta7-positive lymphocytes to oral and skin keratinocytes. Immunology 98:9–18

Butcher EC, Picker LJ (1996) Lymphocyte homing and homeostasis. Science 272:60–66

Camp RL, Scheynius A, Johansson C, Pure E (1993) CD44 is necessary for optimal contact allergic responses but is not required for normal leukocyte extravasation. J Exp Med 178:497–507

Campbell JJ, Haraldsen G, Pan J, Rottman J, Qin S, Ponath P, Andrew DP, Warnke R, Ruffing N, Kassam N, Wu L, Butcher EC (1999) The chemokine receptor CCR4 in vascular recognition by cutaneous but not intestinal memory T cells. Nature 400:776–780

Carlos TM, Harlan JM (1994) Leukocyte-endothelial adhesion molecules. Blood 84:2068–2101

Castells-Rodellas A, Castell JV, Ramirez-Bosca A, Nicolas JF, Valcuende-Cavero F, Thivolet J (1992) Interleukin-6 in normal skin and psoriasis. Acta Derm Venereol 72:165–8

Cepek KL, Shaw SK, Parker CM, Russell GJ, Morrow JS, Rimm DL, Brenner MB (1994) Adhesion between epithelial cells and T lymphocytes mediated by E-cadherin and the alphaEbeta7 integrin. Nature 372:190–193

Christophers E (1996) The immunopathology of psoriasis. Int Arch Allergy Immunol 110:199–206

Condon TP, Flournoy S, Sawyer GJ, Baker BF, Kishimoto TK, Bennett CF (2001) ADAM17 but not ADAM10 mediates tumor necrosis factor-alpha and L-selectin shedding from leukocyte membranes. Antisense Nucleic Acid Drug Dev 11:107–16

Cotran RS, Gimbrone MAJ (1986) Induction and detection of a human endothelial activation antigen in vivo. J Exp Med 164:661–666

Das PK, de Boer OJ, Visser A, Verhagen CE, Bos JD, Pals ST (1994) Differential expression of ICAM-1, E-selectin and VCAM-1 by endothelial cells in psoriasis and contact dermatitis. Acta Derm Venereol Suppl 186:21–2

de Vries IJM, Langeveld-Wildschut EG, van Reijsen FC, Bihari IC, Bruijnzeel-Koomen CAFM, Thepen T (1997) Nonspecific T-cell homing during inflammation in atopic dermatitis: expression of cutaneous lymphocyte-associated antigen and integrin $\alpha E\beta 7$ on skin-infiltrating T cells. J Allergy Clin Immunol 100:694–701

Detmar M, Brown LF, Claffey KP, Yeo KT, Kocher O, Jackman RW, Berse B, Dvorak HF (1994) Overexpression of vascular permeability factor/vascular endothelial growth factor and its receptors in psoriasis. J Exp Med 180:1141–1146

Dustin ML, Rothlein R, Springer T (1986) Induction by IL-1 and interferon gamma: Tissue distribution, biochemistry, and function of a natural adherence molecule (ICAM-1) J Immunol 137:245–250

Dustin ML, Singer KH, Springer T (1988) Adhesion of T lymphoblasts to epidermal keratinocytes is regulated by interferon gamma and is mediated by intercellular adhesion molecule-1. J Exp Med 167:1323–1340

Eedy DJ, Burrows D, Bridges JM, Jones FGC (1990) Clearance of severe psoriasis after allogeneic bone marrow transplantation. Br Med J 300:908

Elder JT, Fisher GJ, Lindquist PB, Bennett GL, Pittelkow MR, Coffey RJ, Ellingsworth L, Derynck R, Voorhees JJ (1989) Overexpression of transforming growth factor a in psoriatic epidermis. Science 243:811–814

Ellis CN, Krueger GG (2001) Treatment of chronic plaque psoriasis by selective targeting of memory effector T lymphocytes. N. Engl J Med 345:248–255

Farber JM (1993) HuMIG: a new human member of the chemokine family of cytokines. Biochem. Biophys. Res Commun 192:223–230

Feigelson SW, Grabovsky V, Winter E, Chen LL, Pepinsky RB, Yednock T, Yablonski D, Lobb R, Alon R (2001) The Src kinase p56(lck) up-regulates VLA-4 integrin affinity. Implications for rapid spontaneous and chemokine-triggered T cell adhesion to VCAM-1 and fibronectin. J Biol Chem 276:13891–901

Feizi T (2001) Carbohydrate ligands for the leukocyte-endothelium adhesion molecules, selectins. Results Probl Cell Differ 33:201–23

Fors BP, Goodarzi K, von Andrian UH (2001) L-selectin shedding is independent of its subsurface structures and topographic distribution. J Immunol 167:3642–51

Fuhlbrigge RC, Kieffer JD, Armerding D, Kupper TS (1997) Cutaneous lymphocyte antigen is a specialized form of PSGL-1 expressed on skin-homing T cells. Nature 389:978–981

Gillitzer R, Ritter U, Spandau U, Goebeler M, Bröcker E-B (1996) Differential expression of GRO-α and IL-8 mRNA in psoriasis: a model for neutrophil migration and accumulation in vivo. J Invest Dermatol 107:778–782

Goebeler M, Toksoy A, Spandau U, Engelhardt E, Bröcker EB, Gillitzer R (1998) The C-X-C chemokine Mig is highly expressed in the papillae of psoriatic lesions. J Pathol 184:89–95

Goodfield M, Hull SM, Holland D, Roberts G, Wood E, Reid S, Cunliffe W (1994) Investigations of the 'active' edge of plaque psoriasis: vascular proliferation precedes changes in epidermal keratin. Br J Dermatol 131:808–13

Gottlieb AB, Chang CK, Posnett DN, Fanell B, Tam JP (1988a) Detection of transforming growth factor a in normal, malignant, and hyperproliferative human keratinocytes. J Exp Med 167:670–675

Gottlieb AB, Luster AD, Posnett DN, Carter DM (1988b) Detection of a gamma interferon-induced protein IP-10 in psoriatic plaques. J Exp Med 168:941–948

Gottlieb JL, Gilleaudeau P, Johnson R, Estes L, Woodworth TG, Gottlieb AB, Krueger JG (1995) Response of psoriasis to a lymphocyte-selective toxin (DAB389 IL-2) suggests a primary immune, but not keratinocyte, pathogenic basis. Nature Med 1:442–447

Grabbe S, Varga G, Beissert S, Steinert M, Pendl G, Seeliger S, Bloch W, Peters T, Schwarz T, Sunderkötter C, Scharffetter-Kochanek K (2002) β2 integrins are required for skin homing of primed T cells but not for priming naive T cells. J Clin Invest 109:183–192

Griffiths CE, Railan D, Gallatin WM, Cooper KD (1995) The ICAM-3/ LFA-1 interaction is critical for epidermal Langerhans cell alloantigen presentation to CD4+ T cells. Br J Dermatol 133:823–829

Griffiths CEM, Voorhees JJ, Nickoloff BJ (1989) Characterization of intercellular adhesion molecule-1 and HLA-DR expression in normal and inflamed skin: Modulation by recombinant gamma interferon and tumor necrosis factor. J Am Acad Dermatol 20:617–629

Groves RW, Allen MH, Barker JN, Haskard DD, MacDonald DM (1991) Endothelial leukocyte adhesion molecule-1 (ELAM-1) expression in cutaneous inflammation. Br J Dermatol 124:117–123

Groves RW, Ross EL, Barker JN, MacDonald DM (1993) Vascular cell adhesion molecule-1 (VCAM-1): expression in normal and diseased skin and regulation in vivo by interferon gamma. J Am Acad Dermatol 29:67–72

Gudmundsdottir AS, Sigmundsdottir H, Sigurgeirsson B, Good MF, Valdimarsson H, Jonsdottir I (1999) Is an epitope on keratin 17 a major target for autoreactive T lymphocytes in psoriasis? Clin Exp Immunol 117:580–586

Hafezi-Moghadam A, Ley K (1999) Relevance of L-selectin shedding for leukocyte rolling in vivo. J Exp Med 189:939–48

Hafezi-Moghadam A, Thomas KL, Prorock AJ, Huo Y, Ley K (2001) L-selectin shedding regulates leukocyte recruitment. J Exp Med 193:863–72

Hemler ME (1990) VLA proteins in the integrin family: Structures, functions, and their role on leukocytes. Annu Rev Immunol 8:365–400

Hertle MD, Kubler MD, Leigh IM, Watt FM (1992) Aberrant integrin expression during epidermal wound healing and in psoriatic epidermis. J Clin Invest 89:1982–1901

Homey B, Alenius H, Müller A, Soto H, Bowman EP, Yuan W, McEvoy L, Lauerma AI, Assmann T, Bünemann E, Lehto M, Wolff H, Yen D, Marxhausen H, To W, Sedgwick J, Ruzicka T, Lehmann P, Zlotnik A (2002) CCL27-CCR10 interactions regulate T cell-mediated skin inflammation. Nat Med 8:157–165

Homey B, Dieu-Nosjean MC, Wiesenborn A, Massacrier C, Pin JJ, Oldham E, Catron D, Buchanan ME, Müller A, deWaal Malefyt R, Deng G, Orozco R, Ruzicka T, Lehmann P, Lebecque S, Caux C, Zlotnik A (2000) Up-regulation of macrophage inflammatory protein-3α/CCL20 and CC chemokine receptor 6 in psoriasis. J Immunol 164:6621–6632

Homey B, Wang W, Soto H, Buchanan ME, Wiesenborn A, Catron D, Müller A, McClanahan TK, Dieu-Nosjean MC, Orozco R, Ruzicka T, Lehmann P, Oldham E, Zlotnik A (2000) The orphan chemokine receptor G protein-coupled receptor-2 (GPR-2, CCR10) binds the skin-associated chemokine CCL27 (CTACK/ALP/ILC). J Immunol 164:3465–3470

Hong K, Chu A, Ludviksson BR, Berg EL, Ehrhardt RO (1999) IL-12, independently of IFNgamma, plays a crucial role in the pathogenesis of a murine psoriasis-like skin disorder. J Immunol 162:7480–7491

Hwang ST (2001) Mechanisms of T cell homing to skin. Adv Dermatol 17:211–241

Hynes RO (1992) Integrins: Versatility, modulation, and signaling in cell adhesion. Cell 69:11–25

Ishida-Yamamoto A, Iizuka H (1995) Differences in involucrin immunolabeling within cornified cell envelopes in normal and psoriatic epidermis. J Invest Dermatol 104:391–395

Issekutz AC, Issekutz TB (2002) The role of E-selectin, P-selectin, and very late activation antigen-4 in T lymphocyte migration to dermal inflammation. J Immunol 168 1934–9

Karecla PI, Bowden SJ, Green SJ, Kilshaw PJ (1995) Recognition of E-cadherin on epithelial cells by the mucosal T cell integrin alphaM290beta7 (alphaEbeta7). Eur J Immunol 25:852–856

Kashihara-Sawami M, Norris DA (1992) The state of differentiation of cultured human keratinocytes determines the level of intercellular adhesion molecule-1 (ICAM-1) expression induced by gamma interferon. J Invest Dermatol 98:852–856

Kellner J, Konter U, Sterry W (1992) Overexpresion of ECM-receptors (VLA-3, 5 and 6) on psoriatic keratinocytes. Br J Dermatol 125:211–215

Kilshaw PJ, Baker KC (1988) A unique surface antigen on intraepithelial lymphocytes in the mouse. Immunol Lett 18:149–54

Konter U, Kellner I, Sterry W (1989) Adhesion molecule mapping in normal human skin. Arch Dermatol. Res 281:454–462

Krueger GG (2002) Selective targeting of T cell subsets: focus on alefacept – a remittive therapy for psoriasis. Expert Opin Biol Ther 2:431–441

Kulke R, Tödt-Pingel I, Rademacher D, Röwert J, Schröder JM, Christophers E (1996) Co-localized overexpression of GRO-α and IL-8 mRNA is restricted to the suprapapillary layers of psoriatic lesions. J Invest Dermatol 106:526–530

Kulke R, Bornscheuer E, Schlüter C, Bartels J, Röwert J, Sticherling M, Christophers E (1998) The CXC receptor 2 is overexpressed in psoriatic epidermis. J Invest Dermatol 110:90–94

Kunkel EJ, Butcher EC (2002) Chemokines and the tissue-specific migration of lymphocytes. Immunity 16:1–4

Kunstfeld R, Lechleitner S, Groger M, Wolff K, Petzelbauer P (1997) HECA-452+ T cells migrate through superficial vascular plexus but not through deep vascular plexus endothelium. J Invest Dermatol 108:343–348

Kupper TS (1990) Immune and inflammatory processes in cutaneous tissues. Mechanisms and speculations. J Clin Invest 86:1783–1786

Leung DY, Gately MK, Trumble A, Ferguson-Darnell B, Schlievert PM, Picker LJ (1995a) Bacterial superantigens induce T-cell expression of the skin-selective homing receptor, the cutaneous lymphocyte-associated antigen, via stimulation of interleukin 12 production. J Exp Med 181:747–753

Leung DYM, Travers JB, Giorno R, Norris DA, Skinner R, Aelion J, Kazemi LV, Kim MH, Trumble AE, Kotb M, Schlievert PM (1995b) Evidence for a streptococcal superantigen-driven process in acute guttate psoriasis. J Clin Invest 96:2106–2112

Lewis HM, Baker BS, Bokth S, Powles AV, Garioch JJ, Valdimarsson J, Fry L (1993) Restricted T-cell receptor Vbeta gene usage in the skin of patients with guttate and chronic plaque psoriasis. Br J Dermatol 129:514–520

Ley K (2001) Functions of selectins. Results Probl Cell Differ 33:177–200

Liao F, Rabin RL, Yannelli JR, Koniaris LG, Vanguri P, Farber JM (1995) Human Mig chemokine: biochemical and functional characterization. J Exp Med 182:1301–1314

Mazanet MM, Neote K, Hughes CCW (2000) Expression of IFN-inducible T cell a chemoattractant by human endothelial cells is cyclosporin A-resistant and promotes T cell adhesion: implications for cyclospori A-resistant immune inflammation. J Immunol 164:5383–5388

McEver RP, Beckstead JH (1989) GMP-140, a platelet alpha granule membrane protein, is also synthesized by vascular endothelial cells and is localized in Waibel-Palade bodies. J Clin Invest 84:92–99

Menssen A, Trommler P, Vollmer S, Schendel D, Albert E, Gurtler L, Riethmuller G, Prinz JC (1995) Evidence for an antigen-specific cellular immune response in skin lesions of patients with psoriasis vulgaris. J Immunol 155:4078–4083

Michel G, Auer H, Kemeny L, Böcking A, Ruzicka T (1996) Antioncogene p53 and mitogenic cytokine interleukin-8 aberrantly expressed in psoriatic skin are inversely regulated by the antipsoriatic drug tacrolimus (FK506). Biochem Pharmacol 51:1315–1320

Michel G, Mirmohammadsadegh A, Olasz E, Jarzebska-Deussen B, Müschen A, Kemeny L, Abts HF, Ruzicka T (1997) Demonstration and functional analysis of IL-10 receptors in human epidermal cells: decreased expression in psoriatic skin, down-modulation by IL-8, and upregulation by an antipsoriatic glucocorticosteroid in normal cultured keratinocytes. J Immunol 159:6291–6297

Mueller W, Herrmann B (1979) Cyclosporin A for psoriasis. N Engl J Med 301:555

Neuner P, Urbanski A, Trautinger F, Moller A, Kirnbauer R, Kapp A, Schopf E, Schwarz T, Luger TA (1991) Increased IL-6 production by monocytes and keratinocytes in patients with psoriasis. J Invest Dermatol 97:27–33

Nickoloff BJ (1991) The cytokine network of psoriasis. Arch Dermatol 127:871–884

Nickoloff BJ (1999) The immunologic and genetic basis of psoriasis. Arch Dermatol 135:1104–1110

Nickoloff BJ (1988) Role of interferon-gamma in cutaneous trafficking of lymphocytes with emphasis on molecular and cellular adhesion events. Arch Dermatol 124:1835–1845

Nickoloff BJ, Griffiths CE, Barker JN (1990) The role of adhesion molecules, chemotactic factors, and cytokines in inflammatory and neoplastic skin disease–1990 update. J Invest Dermatol 94:151S-157S

Nicolas JF, Chamchick N, Thivolet J, Wijdenes J, Morel P, Revillard JP (1991) CD4 antibody treatment of severe psoriasis. Lancet 338:321

Nicolas JF, Rizova H, Demidem A, Thivolet J, Morel P, Revillard JP (1992) CD4 antibody therapy and cyclosporin A differentially affect HLA-DR and ICAM-1 expression in psoriatic skin. J Invest Dermatol 98:943–944

Olivry T, Moore PF, Naydan DK, Danilenko DM, Affolter VK (1995) Investigation of epidermotropism in canine mycosis fungoides: Expression of intercellular adhesion molecule-1 (ICAM-1) and beta-2 integrins. Arch Dermatol Res 287:186–192

Parker CM, Cepek KL, Russell GJ, Shaw SK, Posnett DN, Schwarting R, Brenner MB (1992) A family of beta 7 integrins on human mucosal lymphocytes. Proc Natl Acad Sci USA 89 1924–1929

Pauls K, Schön M, Kubitza RC, Homey B, Wiesenborn A, Lehmann P, Ruzicka T, Parker CM, Schön MP (2001) Role of integrin $\alpha E(CD103)/\beta7$ for tissue-specific epidermal localization of CD8+ T lymphocytes. J Invest Dermatol 117:569–575

Petzelbauer P, Pober JS, Keh A, Braverman IM (1994) Inducibility and expression of microvascular endothelial adhesion molecules in lesional, perilesional, and uninvolved skin of psoriatic patients. J Invest Dermatol 103:300–305

Picker LJ, Kishimoto TK, Smith CW, Warnock RA, Butcher EC (1991) ELAM-1 is an adhesion molecule for skin-homing T-cells. Nature 349:796–799

Picker LJ, Michie SA, Rott LS, Butcher EC (1990) A unique phenotype of skin-associated lymphocytes in humans. Preferential expression of the HECA-425 epitope by benign and malignant T cells at cutaneous sites. Am J Pathol 136:1053–1068

Pinkus H, Mehregan AH (1966) The primary histologic lesion of seborrhoeic dermatitis and psoriasis. J Invest Dermatol 46:109–116

Prinz J, Braun-Falco O, Meurer M, Daddona P, Reiter C, Rieber P, Riethmuller G (1991) Chimaeric CD4 monoclonal antibody in treatment of generalized pustular psoriasis. Lancet 338:320–321

Prinz JC, Gross B, Vollmer S, Trommler P, Strobel I, Meurer M, Plewig G (1994) T cell clones from psoriasis skin lesions can promote keratinocyte proliferation in vitro via secreted products. Eur J Immunol 24:593–598

Ramirez-Bosca A, Martinez-Ojeda L, Valcuende-Cavero F, Castells-Rodellas A (1988) A study of local immunity in psoriasis. Br J Dermatol 119:587–595

Raychaudhuri SP, Jiang WY, Farber EM, Schall TJ, Ruff MR, Pert CB (1999) Upregulation of RANTES in psoriatic keratinocytes: a possible pathogenic mechanism for psoriasis. Acta Derm Venereol 79:9–11

Reinhardt PH, Elliott JF, Kubes P (1997) Neutrophils can adhere via alpha4beta1-integrin under flow conditions. Blood 89:3837–3846

Robert C, Kupper TS (1999) Inflammatory skin diseases, T cells, and immune surveillance. N. Engl J Med 341:1817–1828

Rothe MJ, Nowak M, Kerdel FA (1990) The mast cell in health and disease. J Am Acad Dermatol 23:615–624

Rottman JB, Smith TL, Ganley KG, Kikuchi T, Krueger JG (2001) Potential role of the chemokine receptors CXCR3, CCR4, and the integrin $\alpha E\beta7$ in the pathogenesis of psoriasis vulgaris. Lab Invest 81:335–347

Schechner JS, Edelson RL, McNiff JM, Heald PW, Pober JS (1999) Integrins alpha4beta7 and alphaEbeta7 are expressed on epidermotropic T cells in cutaneous T cell lymphoma and spongiotic dermatitis. Lab Invest 79:601–607

Schlaak JF, Buslau M, Jochum W, Hermann E, Girndt M, Gallati H, Meyer zum Büschenfelde KH, Fleischer B (1994) T cells involved in psoriasis vulgaris belong to the Th1 subset. J Invest Dermatol 102:145–149

Schön MP, Arya A, Murphy EA, Adams CM, Strauch UG, Agace WW, Marsal J, Donohue JP, Her H, Beier DR, Olson S, Lefrancois L, Brenner MB, Grusby MJ, Parker CM (1999) Mucosal T lymphocyte numbers are selectively reduced in integrin a_E (CD103) deficient mice. J Immunol 162:6641–6649

Schön MP, Detmar M, Parker CM (1997) Murine psoriasis-like disorder induced by naive CD4[+] T-cells. Nature Med 3:183–188

Schön MP, Krahn T, Schön M, Rodriguez ML, Antonicek H, Schultz JE, Ludwig RJ, Zollner TM, Bischoff E, Bremm KD, Schramm M, Henninger K, Kaufmann R, Gollnick HPM, Parker CM, Boehncke WH (2002) Efomycine M, a new specific inhibitor of selectin, impairs leukocyte adhesion and alleviates cutaneous inflammation. Nat Med 8:366–372

Schön MP, Ruzicka T (2001) Psoriasis: The plot thickens... Nature Immunol 2:91

Shieh CC, Sadasivan BK, Russell GJ, Schön MP, Parker CM, Brenner MB (1999) Lymphocyte adhesion to epithelia and endothelia mediated by the LEEP-CAM glycoprotein. J Immunol 163:1592–1601

Shimizu Y, Newman W, Gopal TV, Horgan KJ, Graber N, Beall LD, van Seventer GA, Shaw S (1991) Four molecular pathways of T cell adhesion to endothelial cells: roles of LFA-1, VCAM-1, and ELAM-1 and changes in pathway hierarchy under different activation conditions. J Cell Biol 113:1203–1212

Simonitsch I, Volc-Platzer B, Mosberger I, Radaszkiewicz T (1994) Expression of monoclonal antibody defined $aE\beta7$ integrin in cutaneous T cell lymphoma. Am J Pathol 145:1148–1158

Singbartl K, Thatte J, Smith ML, Wethmar K, Day K, Ley K (2001) A CD2-green fluorescence protein-transgenic mouse reveals very late antigen-4-dependent CD8+ lymphocyte rolling in inflamed venules. J Immunol 166:7520–6

Skov L, Chan LS, Fox DA, Larsen JK, Voorhees JJ, Cooper KD, Baadsgaard O (1997) Lesional psoriatic T cells contain the capacity to induce a T cell activation molecule CDw60 on normal keratinocytes. Am J Pathol 150:675–683

Smith CH, Barker JNWN, Morris RW, MacDonald DM, Lee TH (1993) Neuropeptides induce rapid expression of endothelial cell adhesion molecules and elicit granulocytic infiltration in human skin. J Immunol 151:3274–3282

Springer TA (1994) Traffic signals for lymphocyte recirculation and leukocyte emigration: the multistep paradigm. Cell 76:301–314

Sterry W, Mielke V, Konter U, Kellner I, Boehncke WH (1992) Role of β1 integrins in epidermotropism of malignant T cells. Am J Pathol 141:855–860

Stoler A, Kopan R, Duvic M, Fuchs E (1988) Use of monospecific antisera and cRNA probes to localize the major changes in keratin expression during normal and abnormal epidermal differentiation. J Cell Biol 107:427–446

Strange P, Cooper KD, Hansen ER, Fisher GJ, Larsen JK, Fox D, Krag C, Voorhees JJ, Baadsgaard O (1993) T-lymphocyte clones initiated from lesional psoriatic skin release growth factors that induce keratinocyte proliferation. J Invest Dermatol 101:695–700

Strauch UG, Mueller RC, Li XY, Cernadas M, Higgins JMC, Binion DG, Parker CM (2001) Integrin αE(CD103)β7 mediates adhesion to intestinal microvascular endothelial cell lines via an E-cadherin-independent interaction. J Immunol 166:3506–3514

Todderud G, Nair X, Lee D, Alford J, Davern L, Stanley P, Bachand C, Lapointe P, Marinier A, Menard M, Wright JJ, Bajorath J, Hollenbaugh D, Aruffo A, Tramposch KM (1997) BMS-190394, a selectin inhibitor, prevents rat cutaneous inflammatory reactions. J Pharmacol Exp Ther 282:1298–1304

Tomfohrde J, Silverman A, Barnes R, Fernandez-Vina MA, Young M, Lory D, Morris L, Wuepper KD, Stastny P, Menter A, Bowcock A (1994) Gene for familial psoriasis susceptibility mapped to the distal end of human chromosome 17q. Science 264:1141–1145

Travers JB, Hamid QA, Norris DA, Kuhn C, Giorno RC, Schlievert PM, Farmer ER, Leung DY (1999) Epidermal HLA-DR and the enhancement of cutaneous reactivity to superantigenic toxins in psoriasis. J Clin Invest 104:1181–1189

Valdimarsson H, Sigmundsdottir H, Jonsdottir H (1997) Is psoriasis induced by streptococcal superantigens and maintained by M-protein-specific T cells that cross-react with keratin? Clin Exp Immunol 107:21s–24s

van de Kerkhof PC, Goos M, Christophers E, Baudin M, Dupuy P (1995) Inhibitor of the release of mast cell mediators does not improve the psoriatic plaque. Skin Pharmacol 8:25–9

Varki A (1994) Selectin ligands. Proc Natl Acad Sci USA 91:7390–7397

Veale D, Rogers S, Fitzgerald O (1995) Immunolocalization of adhesion molecules in psoriatic arthritis, psoriatic and normal skin. Br J Dermatol 132:32–8

Vekony MA, Holder JE, Lee AJ, Horrocks C, Eperon IC, Camp RD (1997) Selective amplifications of T-cell receptor variable region species is demonstrable but not essential in early lesions of psoriasis vulgaris: analysis by anchored polymerase chain reaction and hypervariable region size spectratyping. J Invest Dermatol 109:5–13

Vestergaard C, Gesser B, Lohse N, Jensen SL, Sindet-Pedersen S, Thestrup-Pedersen K, Matsushima K, Larsen CG (1997) Monocyte chemotactic and activating factor (MCAF/MCP-1) has an autoinductive effect in monocytes, a process regulated by IL-10. J Dermatol Sci 15:14–22

von Andrian UH, Chambers JD, McEvoy LM, Bargatze RF, Arfors KE, Butcher EC (1991) Two-step model of leukocyte-endothelial cell interaction in inflammation: distinct roles for LECAM-1 and the leukocyte beta-2 integrins in vivo. Proc Natl Acad Sci USA 88:7538–7542

von Andrian UH, Mackay CR (2000) T-cell function and migration. Two sides of the same coin. N Engl J Med 343:1020–1034

Wagner LA, Brown T, Gil S, Frank I, Carter W, Tamura R, Wayner EA (1999) The keratinocyte-derived cytokine IL-7 increases adhesion of the epidermal T cell subset to the skin basement membrane protein laminin-5. Eur J Immunol 29:2530–2538

Walton LJ, Thornhill MH, Macey MG, Farthing PM (1997) Cutaneous lymphocyte associated antigen (CLA) and $\alpha E\beta 7$ integrins are expressed by mononuclear cells in skin and oral lichen planus. J Oral Pathol Med 26:402–407

Wayner EA, Gil SG, Murphy GF, Wilke MS, Carter WG (1993) Epiligrin, a component of epithelial basement membranes, is an adhesive ligand for a3b1 positive T lymphocytes. J Cell Biol 121:1141–1152

Weiss RA, Eichner R, Sun TT (1984) Monoclonal antibody analysis of keratin expression in epidermal diseases: A 48- and a 56-kD keratin as molecular markers for hyperproliferative keratinocytes. J Cell Biol 98:1397–1406

Williams IR, Kupper TS (1994) Epidermal expression of intercellular adhesion molecule-1 is not a primary inducer of cutaneous inflammation in transgenic mice. Proc Natl Acad Sci USA 91:9710–9714

Wrone-Smith T, Nickoloff BJ (1996) Dermal injection of immunocytes induces psoriasis. J Clin Invest 98:1878–1887

Yamamoto T, Eckes B, Mauch C, Hartmann K, Krieg T (2000) Monocyte chemoattractant protein-1 enhances gene expression and synthesis of matrix metalloproteinase-1 in human fibroblasts by an autocrine IL-1α loop. J Immunol 164:6174–6179

Zhao LC, Edgar JB, Dailey MO (2001) Characterization of the rapid proteolytic shedding of murine L-selectin. Dev Immunol 8:267–77

Zlotnik A, Morales J, Hedrick JA (1999) Recent advances in chemokines and chemokine receptors. Crit Rev Immunol 19:1–47

Zollner TM, Podda M, Pien C, Elliott PJ, Kaufmann R, Boehncke WH (2002) Proteasome inhibition reduces superantigen-mediated T cell activation and the severity of psoriasis in a SCID-hu model. J Clin Invest 109:671–679

4 The Role of Fucosylation in Leukocyte Adhesion Deficiency II

D. Vestweber, K. Lühn, T. Marquardt, M. Wild

4.1 The Pathophysiology of LAD II

Two leukocyte adhesion deficiencies (LAD) have been described in humans: LAD I and LAD II. Each disease blocks a different step in the sequence of leukocyte–endothelial cell interactions, which is generally referred to as the multistep adhesion cascade (Butcher and Picker 1996; Springer 1994). The cascade is initiated by the tethering and rolling of leukocytes on activated endothelial cells. This first step is mediated by the interaction of selectins (two on endothelium and one on leukocytes) with their glycoprotein ligands (Vestweber and Blanks 1999). Subsequently, activation of leukocytes through the action of chemokines leads to the transformation of leukocyte integrins into a high-affinity state. The integrins are now able to arrest the leukocytes, which then migrate to sites where they can finally extravasate into inflamed tissue surrounding the activated blood vessel. Any interference with these interactions should therefore reduce the emigration of leukocytes to sites of infection, thus compromising immune responses.

Indeed, it has been found that the integrin-mediated firm adhesion of leukocytes to endothelium is inhibited in LAD I, resulting in an immunodeficiency (Hayward et al. 1979; von Andrian et al. 1993). LAD I is caused by mutations in the gene coding for the $\beta 2$ integrin subunit (CD18; Anderson and Springer 1987; Hogg et al. 1999; Kuijpers et al. 1997b) leading to deficient expression or dysfunction of $\beta 2$-containing integrins. CD18 is also important in the activation of T and B cells, a fact that contributes to the severity of the immunodeficiency seen in LAD I.

The second adhesion deficiency, LAD II (also termed congenital disorder of glycosylation-IIc), was first described in 1992 in two children of Arab origin (Etzioni et al. 1992; Frydman et al. 1992). LAD II is a very rare congenital disease (six cases worldwide) which interrupts the adhesion cascade at an earlier step than LAD I. Here it is the first step, the tethering and rolling, which is prevented, again causing an immunodeficiency. In LAD II, the selectin ligands appeared to be defective. Selectins bind to glycoproteins or glycolipids which are decorated with fucosylated carbohydrate structures identical or similar to sialyl-Lewis^x (sLe^x) (NeuAcα2-3Galβ1-4 (Fucα1-3)GlcNAc). On LAD II leukocytes, a nearly complete deficiency of sLe^x expression was observed (Etzioni et al. 1992), accounting for the inability of the leukocytes to roll on endothelium (von Andrian et al. 1993).

In addition to the strongly reduced expression of sLe^x, deficiencies in the formation of other oligosaccharides were found in LAD II patients. Thus, neither the H blood group antigen (Bombay phenotype) nor the Lewis blood group antigens Le^a and Le^b are expressed. Furthermore, there is a defect in the expression of secretor blood group antigens (nonsecretor phenotype; Frydman et al. 1992; Shechter et al. 1995). The fact that all these antigens are fucosylated oligosaccharides and carry fucose in different linkages implies a general defect in fucosylation as the biochemical basis of the disease. Until recently, the genetic defect causing the hypofucosylation was unknown.

The defect in fucosylation clearly compromises the immune system. During their first years of life, LAD II patients suffer from recurrent infections so that at least in one case (Marquardt et al. 1999a) permanent prophylactic treatment with antibiotics was necessary. Later in life, the immune defect can become milder, with periodontitis being the major persistent manifestation, as found in some LAD II patients (Etzioni et al. 1998).

On a cellular level, the immune defect in LAD II has mainly been analyzed in neutrophils. In in vitro assays, these leukocytes failed to interact with E- and P-selectin (Marquardt et al. 1999 b; Phillips et al. 1995), whereas their opsonophagocytic function was normal (Etzioni et al. 1992). Intravital microscopy experiments with patients' neutrophils in inflamed rabbit venules showed an 80% reduction of the rolling fraction of neutrophils and an increase of rolling velocity of residual rolling cells (von Andrian et al. 1993). Induction of firm adhesion in the presence of the chemoattractant leukotriene B4 was also severely impaired, as could be expected for cells that fail to roll. Interestingly, when vascular flow was stopped, LAD II neutrophils became attached and numerous cells remained so when flow was re-established (von Andrian et al. 1993). Under these conditions, some neutrophils were able to emigrate from the venules. However, skin window and skin chamber tests performed on LAD II patients (i.e., under normal flow conditions) showed a strong reduction of neutrophil and monocyte migration to the skin (Price et al. 1994). Taken together, these results show that under conditions of physiological shear forces, firm adhesion and transmigration of neutrophils are dependent on the preceding rolling step. This appears to be the basis of the immunodeficiency seen in LAD II in which rolling is defective. It has been proposed that the experimental conditions of stopped flow and the subsequent ability of neutrophils to transmigrate may have a physiological correlate in cases of edema and vascular stasis at inflammatory sites (Becker and Lowe 1999). This may in part explain why the immunodeficiency in LAD II is relatively mild in comparison to LAD I.

LAD II is characterized by significantly increased peripheral neutrophil counts. The cause of this phenomenon is not clear. Reduced extravasation of neutrophils is an obvious explanation. Another possible cause of the leukocytosis is provided by the finding that in one patient (Y. W.) marrow neutrophil production was increased eightfold (Price et al. 1994). With the neutrophils being unable to extravasate, one would expect an increased survival time of these cells in the vasculature. However, the neutrophil half-life in patient Y.W. appeared to be less than half of normal values (Price et al. 1994). This implies that reduced extravasation and/or increased neutrophil production overcompensate the reduced survival of these cells.

The above-mentioned neutrophil defects in LAD II, including impaired rolling, reduced migration to sites of inflammation, and neutrophilia are in agreement with results obtained in P- and/or E-selectin null mice and mice deficient in fucosyltransferase VII, the enzyme which is primarily responsible for fucosylation of glycoconjugates in neutrophils (Etzioni et al. 1999; Maly et al. 1996).

Lymphocyte function and migration in LAD II have also been studied. Total lymphocyte numbers appeared to be normal in LAD II patients with the exception of one patient (Marquardt et al. 1999a) in whom the number of lymphocytes was increased. The subset composition of T lymphocytes was normal in all patients and activation of T-, B-, and NK-cells using various stimuli also appeared to be unaffected (Etzioni et al. 1992; Kuijpers et al. 1997a; Price et al. 1994). The only lymphocyte defect that was observed was a strong reduction in delayed type hypersensitivity (DTH) reactions (Kuijpers et al. 1997a). Since DTH reactions are largely mediated by the release of cytokines by type I helper T cells (T_H1), this finding indicates that in LAD II patients T_H1 cells either poorly migrate to the skin or that their effector functions are suppressed. This is in agreement with studies in mice that showed that the migration of T_H1 cells to inflamed skin is dependent on P- and E-selectin and the fucosylated selectin ligand P-selectin glycoprotein ligand-1 (PSGL-1; Austrup et al. 1997; Borges et al. 1997). Lymphocyte homing to lymphoid tissues is another important issue, but this has not been experimentally addressed in LAD II.

A hallmark of LAD II is the combination of an immunodeficiency with developmental abnormalities (Etzioni et al. 1992; Frydman et al. 1992; Marquardt et al. 1999a), the latter not being found in LAD I. The most severe developmental defects are found in the patients' psychomotor and mental capabilities. The abilities for directional movement, sitting, and walking are strongly delayed. Speech development is also severely retarded. These defects indicate a strong influence of fucosylation on brain development, which is also underlined by the microcephaly and cortical atrophy that have been described for LAD II. These neurodevelopmental defects are accompanied by growth retardation and clinical stigmata like a depressed nasal bridge and crowded toes.

The molecular basis for the effect of hypofucosylation on neurodevelopment and growth is not known. In analogy to selectin ligands, fucosylated glycoconjugates that participate in development may also be expressed on cell surfaces functioning in cell–cell interactions. O-fucosylation has been recently shown to play essential roles in embryonal development of *Drosophila* and mammals. Fringe was detected as a mutant that affects the function of the receptor Notch during inductive signaling processes in the wing, eye, leg, and ovary (reviewed in Irvine 1999). The Fringe protein was found to be a β1,3N-acetyl-glucosaminyltransferase that elongates O-linked fucose residues attached to epidermal growth factor (EGF) repeats (Bruckner et al. 2000; Moloney et al. 2000). And an O-fucosyltransferase (OFUT1) was recently cloned that attaches O-fucose to EGF repeats in a few proteins, including Notch. Inhibition of the expression of OFUT1 in *Drosophila* was recently shown to affect Notch signaling (Okajima and Irvine 2002). These examples highlight the importance and the potential of fucosylation for the process of embryonal development. The molecular basis of the function of fucose in the development of higher order neurological functions is not yet known.

4.2 Development of a Therapy for LAD II

The third LAD II patient was described in 1999 (Marquardt et al. 1999a). This patient (A.C.) was of Turkish descent and showed a more pronounced immunodeficiency and an earlier onset of growth retardation as compared to the two patients initially described. In order to determine whether hypofucosylation in cells of this patient could be corrected by exogenous fucose, fibroblasts from the patient were cultured in the presence of 1 mM or 10 mM L-fucose. After several days of incubation, cells were analyzed and were found to express normal levels of fucosylated glycoconjugates on the cell surface (Marquardt et al. 1999b). Based on this result it was decided to treat the LAD II patient with a fucose-containing diet.

The administration of oral fucose was first tested in healthy volunteers, showed no side effects, and resulted in high fucose levels in serum with a serum half time of 100 min. Fucose supplementation treatment of LAD II patient A.C. was started at 14 months of age by giving 5 doses of fucose per day, starting with 25 mg fucose/kg body weight (BW) per single dose. Over 9 months of treatment supplementation was slowly increased to 5 daily doses with 492 mg/kg BW.

The effect of the treatment on selectin ligand expression in neutrophils was monitored in flow cytometry analyses using the sLe^x-

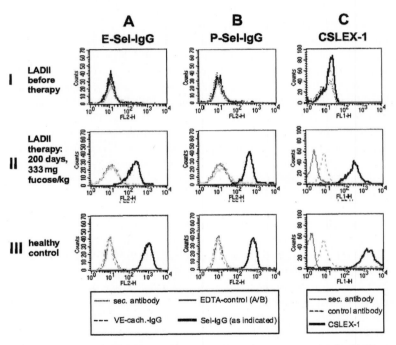

Fig. 1 A–C. Effect of fucose supplementation therapy on selectin binding and sLe^x expression. Neutrophils from a healthy control or LAD II patient A.C. were analyzed by flow cytometry using E-selectin-IgG (*E-Sel-IgG*) or P-selectin-IgG (*P-Sel-IgG*) or antibody CSLEX-1 specific for sLe^x. Histograms with secondary antibody alone, isotype-matched control antibody, or VE-cadherin-IgG construct are shown as negative controls. Binding of selectins in the presence of EDTA is shown as specificity control. (From Marquardt et al. 1999 b)

specific antibody CSLEX-1 as well as E- and P-selectin-IgG fusion proteins (Fig. 1). Before therapy, E- and P-selectin binding was undetectable and CSLEX-1 reactivity was near background (Fig. 1 A–C, I). However, after 200 days of therapy and reaching 333 mg fucose/kg BW per single dose, binding of the endothelial selectins and sLe^x expression were well restored: E-selectin-IgG and CSLEX-1 binding to LAD II cells had reached 20% of that to healthy neutrophils. P-selectin binding was even found to be 70% of normal levels (Fig. 1 A–C, II and III). It was noted that P-selectin binding reappeared at much lower fucose concentrations and lower sLe^x levels than binding of E-selectin, implying that P-selectin binding requires a lower degree of neutrophil surface fucosylation than binding of the other endothelial selectin. Selectins are Ca^{2+}-dependent adhesion molecules. In agreement with this, the presence of etylendiaminetetraacetic acid (EDTA) blocked binding of E- and P-selectin to neutrophils of the treated LAD II patient (Fig. 1 A, B). Further blocking experiments indicated that the restored P-selectin binding under therapy was largely dependent on the interaction of this adhesion molecule with its primary ligand P-selectin glycoprotein ligand-1 (PSGL-1; Marquardt et al. 1999 b).

Since LAD II patients lack the H blood group antigen, it was considered possible that fucose treatment could lead to re-expression of the H-antigen and subsequent autoimmune hemolysis due to a low titer of H-antigen-specific antibodies that was found in the patient. However, the H-antigen was not restored during fucose treatment and no signs of hemolysis were observed (Marquardt et al. 1999 b).

Fucose therapy showed a dramatic effect on the neutrophilia in the LAD II patient. Within the first 10 days of therapy, the highly elevated peripheral neutrophil levels of 10,000 to over 50,000/ll blood returned to the normal range of 1,500–8,500 cells/µl (Fig. 2). The modestly increased levels of blood lymphocytes were not reduced, however, indicating that either the homeostasis of these leukocytes may require a higher degree of fucosylation or that lymphocytes might use exogenous fucose less efficiently than neutrophils.

Importantly, the immune status of the child improved greatly. There were no further infections and antibiotic prophylaxis could be discontinued. The immunodeficiency still appears to be absent after 3 years of treatment.

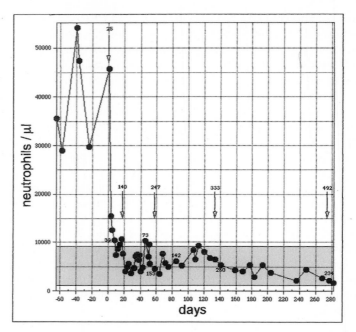

Fig. 2. Effect of fucose therapy on peripheral neutrophil counts. Normal neutrophil numbers are indicated by the *gray bar.* Therapy was started at day 0. Single fucose doses are shown *above the arrows* (in mg/kg body weight). Five doses were administered per day. Serum fucose concentrations (in μM) are given along the graph. (From Marquardt et al. 1999b)

Three months after the start of the therapy, the child's motor control, eye–hand coordination, language, and social interactions were tested and showed significant improvement. This is consistent with the fact that supplemented fucose was able to cross the blood-brain barrier in the patient (Marquardt et al. 1999b), but a causal relationship between the latter two findings cannot be proven. Although the child's development improved, many of the psychomotor defects were not reverted. At the age of four, the patient starts to walk freely and reacts to simple commands, but is still severely retarded.

Following 16 months of therapy, fucose supplementation was interrupted for 9 days due to medical reasons. After 3 days without exogenous fucose, E-selectin binding to neutrophils of patient A.C. was absent, and after 7 days binding of P-selectin and sLe^x-specific antibodies were undetectable (Lühn et al. 2001 a). In addition, the LAD II-typical neutrophilia reappeared, accompanied by fever and

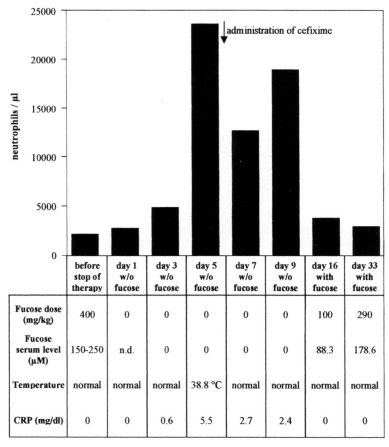

	before stop of therapy	day 1 w/o fucose	day 3 w/o fucose	day 5 w/o fucose	day 7 w/o fucose	day 9 w/o fucose	day 16 with fucose	day 33 with fucose
Fucose dose (mg/kg)	400	0	0	0	0	0	100	290
Fucose serum level (µM)	150-250	n.d.	0	0	0	0	88.3	178.6
Temperature	normal	normal	normal	38.8 °C	normal	normal	normal	normal
CRP (mg/dl)	0	0	0.6	5.5	2.7	2.4	0	0

Fig. 3. Consequences of discontinuation and resumption of fucose therapy. Peripheral neutrophil counts, single fucose doses, serum fucose concentrations, body temperature, and concentrations of C reactive protein (*CRP*) are given for each time point. (From Lühn et al. 2001 a)

elevated levels of C reactive protein (CRP) indicative of inflammatory reactions so that administration of antibiotics was resumed
(Fig. 3). When fucose therapy was re-established later on, neutrophil
counts and CRP concentrations fell to normal levels (Fig. 3) and selectin binding as well as sLe^x expression were restored (Lühn et al.
2001 a). These results provided formal proof that the induction of selectin ligand expression, as well as the reduction of neutrophil
counts and inflammation during therapy, was indeed dependent on
fucose supplementation.

Since the original description of the first two children with LAD
II (Y. W. and S. A.; Etzioni et al. 1992; Frydman et al. 1992), two
more Arab LAD II patients, a boy and a girl, were found by Etzioni
and coworkers. At the age of 4 weeks, fucose supplementation of
these children was started, albeit with considerably lower daily fucose doses than described for the therapy of LAD II patient A. C.
This treatment showed no beneficial effects (Etzioni and Tonetti
2000 a). Treatment was then repeated in the boy (Sturla et al. 2001),
now using the high-dose protocol described in Marquardt et al.
(1999 b). However, again this therapy showed no effect after
150 days of treatment.

Given that the biodistribution of oral fucose is likely to be similar
in patient A. C. and the male patient treated by Etzioni et al., the different outcome of the same treatment indicates that the biochemical
defect in the latter patient is somewhat different, thus preventing external fucose from being utilized for fucosylation. An interesting test
– which has yet to be performed – is whether in vitro fucose treatment of cells of the unsuccessfully treated patient is as inefficient as
the therapy itself. This would help to answer the question whether
the easy and quick in vitro test allows to predict the outcome of fucose supplementation therapy. In this respect it has to be noted that
in in vitro assays exogenous fucose rescued fucosylation not only in
cells of LAD II patient A. C., but also in cells from one of the two
originally described patients (Karsan et al. 1998); however, this patient was not treated because he was above 3 years of age when the
immune status was nearly normal and developmental defects were
more unlikely to improve (Etzioni and Tonetti 2000 b).

Recently, a new LAD II patient of Brazilian origin, the sixth case
worldwide, was found. Treating this patient with oral fucose success-

fully corrected the immunodeficiency similar to the first successfully treated patient (A.C.), further demonstrating the usefulness of this therapy (Hidalgo et al. 2003). Surprisingly, however, this patient seemed to re-express some level of the H-antigen upon fucose therapy, although no signs of hemolysis were detected (Hidalgo et al. 2003).

4.3 Molecular Basis of the Disease

LAD II patients suffer from a generalized fucosylation defect. The defect is seen in several cell types and affects $\alpha1,2$-, $\alpha1,3$-, $\alpha1,4$-, and $\alpha1,6$-linkages of fucose in glycoconjugates (Etzioni et al. 1992; Frydman et al. 1992; Marquardt et al. 1999a). All these fucosylations are performed by different fucosyltransferases and a multiple defect in fucosyltransferase genes seemed unlikely. This notion was confirmed by the finding that activities of $\alpha1,2$- and $\alpha1,3/4$-fucosyltransferases were normal in serum and saliva of LAD II patient Y. W. (Shechter et al. 1995). The fact that exogenous fucose corrected the hypofucosylation in vitro and in vivo excluded the possibility that a common cofactor for fucosyltransferases was defective in LAD II. Thus, the defect had to be found in a step of fucose metabolism preceding the transfer of fucose from its donor GDP-fucose to glycoconjugates.

GDP-fucose is synthesized in the cytosol by two different pathways, a de novo pathway (Ginsburg 1960) which provides up to 90% of the cellular GDP-fucose pool, and a salvage pathway (Yurchenco and Atkinson 1977; Fig. 4). The de novo pathway uses GDP-mannose as substrate, which itself originates from glucose or mannose. The enzymes which are active in this pathway are GDP-mannose-4,6-dehydratase (GMD; Ohyama et al. 1998; Sullivan et al. 1998) and the FX protein which exhibits an epimerase as well as a reductase activity (Chang et al. 1988; Tonetti et al. 1996; Smith et al. 2002). The salvage pathway utilizes fucose that is taken up into the cell or is derived from degraded glycoconjugates. Here, the conversion into GDP-fucose is carried out by a fucose-kinase and a GDP-L-fucose-pyrophosphorylase (Park et al. 1998; Pastuszak et al.

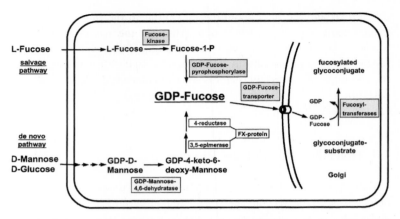

Fig. 4. Pathways of GDP-fucose biosynthesis. GDP-fucose is synthesized from mannose or glucose by the de novo pathway or from fucose by the salvage pathway. GDP-fucose is then transported into the Golgi and utilized by different fucosyltransferases. (From Marquardt et al. 1999b)

1998). GDP-fucose is then transported into the Golgi where it can finally be utilized by the different fucosyltransferases.

Initially it was assumed that LAD II is caused by a defect in GDP-fucose synthesis (Etzioni et al. 1992). The correction of fucosylation in cells of one of the two originally described LAD II patients and in cells from an abortus by external fucose showed that the salvage pathway was functional (Karsan et al. 1998). This was taken as indirect evidence that the de novo pathway is defective in LAD II. Support for this notion came from a report which showed that the activity of the de novo pathway enzyme GMD was affected in cells of one of the first two LAD II patients (Sturla et al. 1998). Here, a reduced maximal activity and a lag phase were observed in in vitro enzyme assays. Curiously however, GMD was expressed at normal levels and no mutation was found when its gene was analyzed in LAD II cells. It was therefore proposed that the genetic defect may lie in a gene coding for a GMD-regulating protein.

A turn in the perception of the molecular defect came with two reports which showed that in LAD II patient A.C. the de novo pathway exhibited normal activity in in vitro assays (Körner et al. 1999) and that on the other hand the import of GDP-fucose into Golgi vesicles was reduced by approximately 80% (Lübke et al. 1999). It was proposed that the reduced import of the nucleotide sugar might be due to a K_m (i.e., affinity) defect in the GDP-fucose transport, implying that strongly increased GDP-fucose levels in the cytosol could be able to drive the transport despite the defect. This would explain why exogenous fucose corrected the fucosylation defect in patient A.C.

Körner et al. (1999) proposed that a transport defect would also give an explanation for the reduced GMD activity that had been observed in cells of the LAD II patient described in Sturla et al. (1998): a reduced transport into the Golgi can be expected to lead to increased cytosolic GDP-fucose concentrations. These, in turn, have been shown to serve as negative feedback signals for GMD (Kornfeld and Ginsburg 1966). It is not clear whether this explanation can be applied to the measurements of GMD activity in which highly diluted lysates with equally diluted cytosolic GDP-fucose were used (Sturla et al. 1998).

A defect in GDP-fucose transport across the Golgi membrane can theoretically originate from a malfunction in one of two proteins, the nucleotide sugar transporter itself or a GDPase which resides in the Golgi membrane and converts GDP into GMP, a product which is required to drive the GDP-fucose import in an antiport system. Lübke et al. (1999), however, could show that the GDPase of patient A.C. had normal activity, leaving the GDP-fucose transporter as the most probable source of the defect.

At the time the GDP-fucose import defect was described no GDP-fucose transporter had been cloned in any species. In order to find such a gene and simultaneously test whether this nucleotide sugar transporter (NST) was defective in LAD II patient A.C., we aimed at transfecting the patient's fibroblasts with cDNAs that showed similarity to known NSTs. We searched the genome of the nematode *Caenorhabditis elegans* for such NST-like genes because it had been fully sequenced and well-analyzed for open reading frames. In addition, it was known that *C. elegans* expresses fucosylated structures and therefore had to contain a GDP-fucose transporter.

We found 16 sequences in the *C. elegans* genome, all of which were likely to encode NST-like multitransmembrane proteins.

Twelve of these sequences were isolated as cDNAs and transiently transfected into the hypofucosylated fibroblasts that had been derived from patient A.C. Only two of these cDNAs showed an influence on fucosylation. One (referred to as ORF-11) had a minor effect, whereas the second (ORF-7) efficiently restored fucosylation in transfected cells, indicating that this cDNA codes for the *C. elegans* GDP-fucose transporter (Lühn et al. 2001 b).

A search of databases for sequences similar to ORF-7 revealed the existence of a human protein of unknown function with 55% amino acid identity with the *C. elegans* GDP-fucose transporter. This putative human orthologue was transiently expressed in LAD II fibroblasts. To detect the few successfully transfected cells, a GFP-fusion protein was cotransfected. Transfected cells acquired the ability to fucosylate glycoconjugates, as was determined by the staining with the fucose-specific lectin from *Aleuria aurantia* (AAL). It was noted that the human protein was considerably more efficient in restoring fucosylation than its *C. elegans* homologue (Lühn et al. 2001 b). These results indicated that the human protein represented the human GDP-fucose transporter and that expression of this protein rescued the defect in LAD II.

As a GDP-fucose transporter, this protein was expected to be located in the Golgi. Indeed, a fusion protein containing the human transporter fused at its carboxy terminus to the green fluorescent protein (GFP) was targeted to the Golgi in transfected COS-7 cells, as determined by its colocalization with the Golgi marker golgin-97. Hydrophobicity analyses suggested that this protein traverses the Golgi membrane ten times (Fig. 5). An even number of 6–10 transmembrane domains with both amino and carboxy termini exposed to the cytosol is typical for nucleotide-sugar transporters (Eckhardt et al. 1999; Hirschberg et al. 1998). The protein was predicted to contain 364 amino acids, which is in good agreement with the molecular weight (39 kDa) of a rat liver Golgi protein that had previously been biochemically characterized and shown to have GDP-fucose transport activity (Puglielli and Hirschberg 1999).

The group that initially described the Golgi import defect also searched for the genetic lesion by expression of a human cDNA library in fibroblasts derived from LAD II patient A.C. (Lübke et al. 2001). They found one cDNA that restored fucosylation in these

Golgi lumen

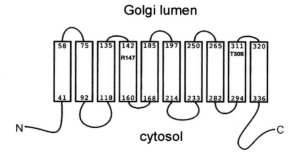

cytosol

Golgi lumen

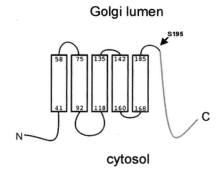

cytosol

Fig. 5. Predicted membrane topology and detected mutations of the GDP-fucose transporter. The topology of the transporter was concluded from two different structure prediction algorithms and comparison with known nucleotide sugar transporters (Lühn et al. 2001 b). Biochemical data from rat Golgi vesicles indicate that the transporter exists as a homodimer (Puglielli and Hirschberg 1999). The *boxes* represent the predicted transmembrane domains. Mutated amino acid residues in patient A.C. (R147) and three other LAD II patients (T308) are depicted in the *upper panel*. The *lower panel* depicts the mutated form of the protein that was found in the most recently identified patient (Hidalgo et al. 2003). This mutated protein results from a single nucleotide deletion leading to an alteration of the open reading frame of the protein after serine 195, introducing 34 random amino acids (shown in *gray*). (Adapted from Lühn et al. 2001 b)

cells and this sequence was identical with the one identified in our experiments. This group verified the GDP-fucose transporter activity of the protein encoded by this cDNA in in vitro transport assays using Golgi-enriched microsomes of transfected LAD II fibroblasts (Lübke et al. 2001).

Both groups sequenced the gene coding for the GDP-fucose transporter in LAD II patient A.C. A point mutation was found causing replacement of arginine 147 by a cysteine (Lübke et al. 2001; Lühn et al. 2001 b). This arginine lies within a conserved stretch of the 4th transmembrane domain (Fig. 5). Moreover, the arginine is conserved in the orthologues in *C. elegans* and *Drosophila melanogaster* and is also found at analogous positions in other nucleotide sugar transporters, pointing to an important role of this particular amino acid (Lühn et al. 2001 b).

The mutated transporter protein was still targeted to the Golgi as shown by the localization of a GFP-fusion protein in COS-7 cells. However, when the mutated GDP-fucose transporter was expressed in LAD II fibroblasts, the protein was not able to restore fucosylation, showing that the R147C mutation functionally inactivates the transporter and represents the genetic defect in LAD II patient A.C.

In addition to this patient, Lübke et al. (2001) analyzed the GDP-fucose transporter mRNAs of two patients (referred to as Y.W. and M.) who had been found by Etzioni and coworkers. In these patients, threonine 308 was replaced by an arginine. Threonine 308 is located in a highly conserved sequence in the ninth transmembrane region (Fig. 5) and is also found in the orthologues in *C. elegans* and *D. melanogaster*. As pointed out by Lübke et al. (2001), arginine 147 as well as threonine 308 lie in transmembrane domains with unusually high hydrophilicity, pointing to a possible role of these amino acids in the gating of the nucleotide sugar. Recently, another LAD II patient was found with a mutation leading to the replacement of threonine 308 (Etzioni et al. 2002; Hirschberg 2001). Thus, four of the five first known LAD II patients have been analyzed and found to display a defect in the GDP-fucose transporter.

The most recently identified new LAD II patient of Brazilian descent, who had been the second patient to be successfully treated with oral fucose (Hidalgo et al. 2003), was shown to have a single nucleotide deletion in the GDP-fucose transporter gene leading to an alteration of the open reading frame of the protein after serine 195, introducing 34 random amino acids followed by a stop codon. This severe mutation results in the truncation of the protein and most likely renders the transporter nonfunctional. If this is true, the success of the treatment with oral L-fucose must be based on an additional transport system, as pointed out by the authors of that report.

4.4 Conclusions and Outlook

An interesting question is whether the three mutations in the GDP-fucose transporter, A147C T308R, and the truncation lead to different biochemical properties of the transporter molecule and may be correlated with the different outcome of fucose therapy. The successful fucose treatment of LAD II patient A.C. indicates that the A147C mutation may reduce the affinity of the NST for GDP-fucose as outlined above. Recently, it was found that a GDP-fucose transporter with the T308R mutation (patient Y.W.) displayed a maximal transport activity (V_{max}) that was only 30% of the normal rate, whereas the affinity was unaffected (Sturla et al. 2001). The LAD II patient who was treated without success (Sturla et al. 2001) also shows the T308R mutation (Etzioni et al. 2002). With a transporter working at a reduced V_{max}, in this child exogenous fucose would not be able to increase the transport rate, explaining the negative outcome of the treatment. In light of the fact that the T308R LAD II patients could not be successfully treated with fucose (Etzioni and Tonetti 2000a; Sturla et al. 2001), it is remarkable that the patient displaying the severely truncated version of the GDP-fucose transporter (Hidalgo et al. 2003) that is most likely not functional could still be successfully treated with fucose. Based on these results, it is not yet understood why the fucose therapy failed for the first patients.

Another open question refers to the fact that Golgi vesicles of LAD II patients show a low residual GDP-fucose transport activity (Lübke et al. 1999; Sturla et al. 2001). One explanation for this is that the mutated transporter molecules themselves may account for the residual transport. Our transfection experiments with the LAD II patient's GDP-fucose transporter do not support such a notion since no increase in fucosylation was detectable despite overexpression of this transporter. However, it is possible that these assays are not sensitive enough to detect very low fucosylation levels. An additional explanation is the possible existence of another, low efficient, transport mechanism. Indeed, we have detected a protein in *C. elegans* encoded by cDNA ORF-11 which restored fucosylation in LAD II cells with low efficiency (Lühn et al. 2001 b). No putative orthologue of this protein has yet been found in humans. Whether ORF-11 is related to the residual GDP-fucose transport activity in LAD II needs to be investigated. Such a second residual transport activity would also explain the success of the fucose therapy in the patient expressing the truncated GDP-fucose transporter.

Finally, the strong neurodevelopmental and growth defects in LAD II point to an important role of fucose in these processes. Patterns of fucosylation have been described in the brain of animals (Yamamoto et al. 1985), and the application of L-fucose has been shown to positively affect learning behavior and protein synthesis in rat brain (Wetzel et al. 1980), but little more is known about the function of fucosylation in the nervous system. One can now make use of the knowledge of the genetic defect in LAD II and generate animals deficient in GDP-fucose transport activity. This may help to provide another piece of the puzzle in our understanding of the relation between development and glycosylation.

References

Anderson DC, Springer TA (1987) Leukocyte adhesion deficiency: an inherited defect in Mac-1, LFA-1, and p150/95 glycoprotein. Ann Rev Med 38:175–192

Austrup K, Vestweber D, Borges E, Löhning M, Bräuer R, Herz U, Renz H, Hallmann R. Scheffold A, Radbruch A, Hamann A (1997) P- and E-selectin mediate recruitment of T-helper-1 but not T-helper-2 cells into inflamed tissues. Nature 385:81–83

Becker DJ, Lowe JB (1999) Leukocyte adhesion deficiency type II. Biochim Biophys Acta 1455:193–204

Borges E, Tietz W, Steegmaier M, Moll, T, Hallmann R, Hamann A, Vestweber D (1997) P-selectin glycoprotein-1 (PSGL-1) on T helper 1 but not T helper 2 cells binds to P-selectin and supports migration into inflamed skin. J Exp Med 573–578

Bruckner K, Perez L, Clausen H, Cohen S (2000) Glycosyltransferase activity of Fringe modulates Notch-Delta interactions. Nature 406:411–415

Butcher EC, Picker LJ (1996) Lymphocyte homing and homeostasis. Science 272:60–66

Chang S, Duerr B, Serif G (1988) An epimerase-reductase in L-fucose synthesis. J Biol Chem 263:1693–1697

Eckhardt M, Gotza B, Gerardy-Schahn R (1999) Membrane topology of the mammalian CMP-sialic acid transporter. J Biol Chem 274:8779–8787

Etzioni A, Frydman M, Pollack S, Avidor I, Phillips ML, Paulson JC, Gershoni-Baruch R (1992) Recurrent severe infections caused by a novel leukocyte adhesion deficiency. N Engl J Med 327:1789–1792

Etzioni A, Gershoni-Baruch R, Pollack S, Shehadeh N (1998) Leukocyte adhesion deficiency type II: long-term follow-up. J Allergy Clin Immunol 102:323–324

Etzioni A, Doerschuk CM, Harlan JM (1999) Of man and mouse: leukocyte and endothelial adhesion molecule deficiencies. Blood 94:3281–3288

Etzioni A, Tonetti M (2000a) Fucose supplementation in leukocyte adhesion deficiency type II. Blood 95:3641–3643

Etzioni A, Tonetti M (2000b) Leukocyte adhesion deficiency II-from A to almost Z. Immunol Rev 178:138–147

Etzioni A, Sturla L, Antonellis A, Green ED, Gershoni-Baruch R, Berninsone PM, Hirschberg CB, Tonetti M (2002) Leukocyte adhesion deficiency (LAD) type II/carbohydrate deficient glycoprotein (CDG) IIc founder effect and genotype/phenotype correlation. Am J Med Genet 110:131–135

Frydman M, Etzioni A, Eidlitz-Markus T, Avidor I, Varsano I, Shechte Y, Orlin JB, Gershoni-Baruch R (1992) Rambam-Hasharon syndrome of psychomotor retardation, short stature, defective neutrophil motility, and Bombay phenotype. Am J Med Genet 44:297–302

Ginsburg V (1960) Formation of guanosine diphosphate L-fucose from guanosine diphosphate D-mannose. J Biol Chem 235:2196–2201

Hayward AR, Leonard J, Wood CBS, Harvey BAM, Greenwood MC, Soohtill JF (1979) Delayed separation of the umbilical cord, wide spread infections, and defective neutrophil mobility. Lancet 1:1099–2002

Hidalgo A, Ma S, Peired AJ, Weiss LA, Cunnigham-Rundles C, Frenette PS (2003) Insights into leukocyte adhesion deficiency type II from a novel mutation in the GDP-fucose transporter gene. Blood 101:1705–1712

Hirschberg CB, Robbins PW, Abeijon C (1998) Transporters of nucleotide sugars, ATP, and nucleotide sulfate in the endoplasmic reticulum and Golgi apparatus. Annu Rev Biochem 67:49–69

Hirschberg CB (2001) Golgi nucleotide sugar transport and leukocyte adhesion deficiency II. J Clin Invest 108:3–6

Hogg N, Stewart MP, Scarth SL, Newton R, Shaw JM, Law SK, Klein N (1999) A novel leukocyte adhesion deficiency caused by expressed but nonfunctional beta2 integrins Mac-1 and LFA-1. J Clin Invest 103:97–106

Irvine KD (1999) Fringe, Notch and making developmental boundaries. Curr Opin Genet Dev 9:434–441

Karsan A, Cornejo CJ, Winn RK, Schwartz BR, Way W, Lannir N, Gershoni-Baruch R, Etzioni A, Ochs HD, Harlan JM (1998) Leukocyte Adhesion Deficiency Type II is a generalized defect of de novo GDP-fucose biosynthesis. Endothelial cell fucosylation is not required for neutrophil rolling on human nonlymphoid endothelium. J Clin Invest 101:2438–2445

Körner C, Linnebank M, Koch HG, Harms E, von Figura K, Marquardt T (1999) Decreased availability of GDP-L-fucose in a patient with LAD II with normal GDP-D-mannose dehydratase and FX protein activities. J Leukoc Biol 66:95–98

Kornfeld RH, Ginsburg V (1966) Control of synthesis of guanosine 5′-diphosphate D-mannose and guanosine 5′-diphosphate L-fucose in bacteria. Biochim Biophys Acta 117:79–87

Kuijpers TW, Etzioni A, Pollack S, Pals ST (1997a) Antigen-specific immune responsiveness and lymphocyte recruitment in leukocyte adhesion deficiency type II. Int Immunol 9:607–613

Kuijpers TW, van Lier R AW, Hamann D, de Boer M, Thung LY, Weening RS, Verhoeven AJ, Roos D (1997b) Leukocyte adhesion deficiency type 1 (LAD-1)/variant. A novel immunodeficiency syndrome characterized by dysfunctional b2 integrins. J Clin Invest 100:1725–1733

Lübke T, Marquardt T, von Figura K, Korner C (1999) A new type of carbohydrate-deficient glycoprotein syndrome due to a decreased import of GDP-fucose into the golgi. J Biol Chem 274:25986–25989

Lübke T, Marquardt T, Etzioni A, Hartmann E, von Figura K, Korner C (2001) Complementation cloning identifies CDG-IIc, a new type of congenital disorders of glycosylation, as a GDP-fucose transporter deficiency. Nat Genet 28:73–76

Lühn K, Marquardt T, Harms E, Vestweber D (2001a) Discontinuation of fucose therapy in LADII causes rapid loss of selectin ligands and rise of leukocyte counts. Blood 97:330–332

Lühn K, Wild MK, Eckhardt M, Gerardy-Schahn R, and Vestweber D (2001b) The gene defective in leukocyte adhesion deficiency II encodes a putative GDP-fucose transporter. Nat Genet 28:69–72

Maly P, Thall AD, Petryniak B, Rogers CE, Smith PL, Marks RM, Kelly RJ, Gersten KM, Cheng G, Saunders TL, Camper SA, Camphausen RT, Sullivan FX, Isogai Y, Hindsgaul O, von Andrian UH, Lowe JB (1996) The α-(1,3)fucosyltransferase Fuc-TVII controls leukocyte trafficking through an essential role in L-, E-, and P-selectin ligand biosynthesis. Cell 86:643–653

Marquardt T, Brune T, Lühn K, Zimmer K-P, Körner C, Fabritz L, van der Werft N, Harms E, von Figura K, Vestweber D, Koch HG (1999a) Leukocyte adhesion deficiency II syndrome, a generalized defect in fucose metabolism. J Pediatr 134:681–688

Marquardt T, Lühn K, Srikrishna G, Freeze HH, Harms E, Vestweber D (1999b) Correction of leukocyte adhesion deficiency Type II with oral fucose. Blood 94:3976–3985

Moloney DJ, Panin VM, Johnston SH, Chen J, Shao L, Wilson R, Wang Y, Stanley P, Orvine KD, Haltiwanger RS, Vogt TF (2000) Fringe is a glycosyltransferase that modifies Notch. Nature 406:369–375

Ohyama C, Smith PL, Angata K, Fukuda MN, Lowe JB, Fukuda M (1998) Molecular cloning and expression of GDP-D-mannose-4,6-dehydratase, a key enzyme for fucose metabolism defective in Lec13 cells. J Biol Chem 273:14582–14587

Okajima T and Irvine KD (2002) Regulation of Notch signaling by O-linked fucose. Cell 111:893–904

Park SH, Pastuszak I, Drake R, Elbein AD (1998) Purification to apparent homogeneity and properties of pig kidney L-fucose kinase. J Biol Chem 273:5685–5691

Pastuszak I, Ketchum C, Hermanson G, Sjoberg EJ, Drake R, Elbein AD (1998) GDP-L-fucose pyrophosphorylase. Purification, cDNA cloning, and properties of the enzyme. J Biol Chem 273:30165–30174

Phillips ML, Schwartz BR, Etzioni A, Bayer R, Ochs HD, Paulson JC, Harlan JM (1995) Neutrophil adhesion in leukocyte adhesion deficiency syndrome type 2. J Clin Invest 96:2898–2906

Price TH, Ochs HD, Gershoni Baruch R, Harlan JM, Etzioni A (1994) In vivo neutrophil and lymphocyte function studies in a patient with leukocyte adhesion deficiency type II. Blood 84:1635–1639

Puglielli L, Hirschberg CB (1999) Reconstitution, identification, and purification of the rat liver golgi membrane GDP-fucose transporter. J Biol Chem 274:35596–35600

Shechter Y, Etzioni A, Levene C, Greenwell P (1995) A Bombay individual lacking H and Le antigens but expressing normal levels of alpha-2- and alpha-4-fucosyltransferases. Transfusion 35:773–776

Smith PL, Myers JT, Rogers CE, Zhou L, Petryniak B, Becker DJ, Homeister JW, Lowe JB. (2002) Conditional control of selectin ligand expression and global fucosylation events in mice with a targeted mutation at the FX locus. J Cell Biol. 158:801–15.

Springer TA (1994) Traffic signals for lymphocyte recirculation and leukocyte emigration: The multistep paradigm. Cell 76:301–314

Sturla L, Etzioni A, Bisso A, Zanardi D, De Flora G, Silengo L, De Flora A, Tonetti M (1998) Defective intracellular activity of GDP-D-mannose-4,6-dehydratase in leukocyte adhesion deficiency type II syndrome. FEBS Lett 429:274–278

Sturla L, Puglielli L, Tonetti M, Berninsone P, Hirschberg CB, De Flora A, Etzioni A (2001) Impairment of the Golgi GDP-L-fucose transport and unresponsiveness to fucose replacement therapy in LAD II patients. Pediatr Res 49:537–542

Sullivan FX, Kumar R, Kriz R, Stahl M, Xu GY, Rouse J, Chang XJ, Boodhoo A, Potvin B, Cumming DA (1998) Molecular cloning of human GDP-mannose 4,6-dehydratase and reconstitution of GDP-fucose biosynthesis in vitro. J Biol Chem 273:8193–8202

Tonetti M, Sturla L, Bisso A, Benatti U, De Flora A (1996) Synthesis of GDP-L-fucose by the human FX protein. J Biol Chem 271:27274–27279

Vestweber D, Blanks JE (1999) Mechanisms that regulate the function of the selectins and their ligands. Physiol Rev 79:181–213

von Andrian UH, Berger EM, Ramezani L, Chambers JD, Ochs HD, Harlan JM, Paulson JC, Etzioni A, Arfors KE (1993) In vivo behavior of neutrophils from two patients with distinct inherited leukocyte adhesion deficiency syndromes. J Clin Invest 91:2893–2897

Wetzel W, Popov N, Lossner B, Schulzeck S, Honza R, Matthies H (1980) Effect of L-fucose on brain protein metabolism and retention of a learned behavior in rats. Pharmacol Biochem Behav 13:765–771

Yamamoto M, Boyer AM, Schwarting GA (1985) Fucose-containing glycolipids are stage- and region-specific antigens in developing embryonic brain of rodents. Proc Natl Acad Sci USA 82:3045–3049

Yurchenco PD, Atkinson PH (1977) Equilibration of fucosyl glycoprotein pools in He La Cells. Biochemistry 16:944–953

5 Substrate Specificity and Synthetic Use of Glycosyltransferases

J. Thiem

5.1 Introduction

Increasingly, glycosyltransferases from all sorts of organisms are being isolated, purified, and well-characterized by biochemists. If they are stable and can be obtained in purer qualities, in principal their application in chemical synthesis as highly selective biocatalysts should be feasible. This could improve and speed up the time-consuming and cumbersome chemical assembly (Paulsen 1990) of the desired, rather complex oligosaccharides or their conjugates. This class of compounds is of particular biochemical and medicinal interest because they play decisive roles in many cellular recognition processes such as cell–cell adhesion events, or as attachment sites for binding of hormones, toxins, viruses, or bacteria (Varki 1993). In addition to their original isolation, more and more cloned glycosyl-transferases became available (Kleene and Berger 1993), and thus

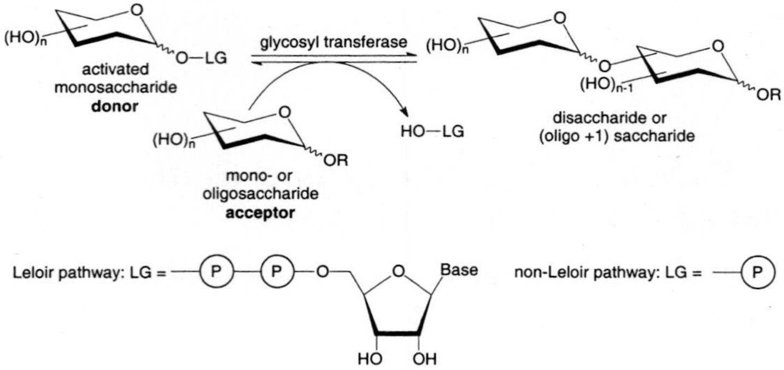

Fig. 1. Schematic presentation of enzymatic glycosylation by glycosyltransferases

the combined chemical–enzymatic synthesis of complex hetero-oligosaccharides is gaining in attraction (Wong et al. 1995).

Glycosyltransferases catalyse the attachment of a monosaccharide via its activated form (the donor substrate) to an accepting mono- or oligosaccharide (the acceptor substrate). In mammalians, oligosaccharides are biosynthesized by the Leloir glycosyltransferase, which requires a nucleotide sugar as the donor (Beyer et al. 1982); however, other "non-Leloir" glycosyltransferases also occur which require a more simple activation as sugar phosphates (Fig. 1).

For the synthetic saccharide chemist, the advantage of the glycosyltransferases lies in the complete regio- and stereocontrol and the almost quantitative yields of the formation of the interglycosidic linkages. Rather disappointing, however, is their sensitivity, which makes their handling generally difficult. Also, they are rare and thus expensive, and this also applies to the activated nucleotide sugars. Further developments are being studied to improve handling, such as immobilization to reduce the costs, e.g., by in situ regeneration of donors, and to widen the possibility by making cloned enzymes available (Fig. 1).

5.2 β1-4-Galactosyltransferase

For years this enzyme (GalT, EC 2.4.1.22/38) was commercially available in larger quantities and thus has been widely studied. One of the early uses for galactosylation was elaborated (Wong et al. 1982) and further improved for preparative purposes (Thiem and Treder 1986) as depicted in Fig. 2. Starting with glucose-6-phosphate (R=OH), enzymatic transfer to the Cori ester (α-D-glucopyranosyl phosphate) and further to uridine-5′-diphosphoglucose (UDP-glucose) results in UDP-galactose, which serves as the donor substrate to allow regio- and stereospecific β-galactosylation of β-N-acetyl glucosaminyl-terminated oligosaccharides as acceptor substrates employing GalT. The released UDP could be transformed

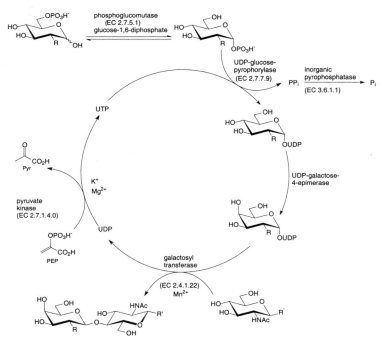

Fig. 2. Cofactor-regeneration cycle for enzymatic galactosylation employing β1-4 galactosyltransferase

into uridine-5′-triphosphate (UTP), with phosphoenol pyruvate catalyzed by pyruvate kinase, thus closing the circle of cofactor regeneration.

In later studies, donor modifications also could be demonstrated and the same reaction could be shown to operate with the 2-deoxy components (R=H). Thus, 2-deoxy-β-galactosylation could be achieved with complex saccharides (Thiem and Wiemann 1991, 1992).

The majority of studies, however, were on modification of the acceptor substrates. The above regeneration method is attractive, yet in the cases where intermediates are readily available one may consider a short cut. This is the case with bovine β1-4GalT and UDP-Glc as activated sugar, and therefore only a two-enzyme system was applied, despite awareness of the fact that UDP may reduce the yields due to feedback inhibition.

One series of experiments was to extend the galactosylation from glucose, the natural substrate for "lactose synthase" (galactosyl transferase plus α-lactalbumin), leading quantitatively to lactose, and test higher cello-oligosaccharides as acceptor substrates. Indeed, cellobiose could be terminally galactosylated to give the trisaccharide in 23% yield. All attempts to employ higher cello-oligomers showed no further transfer. Apparently, this system is governed by a rather firm structural requirement of the acceptor molecule (Fig. 3).

Corresponding experiments were performed with the series of chito-oligosaccharides from GlcNAc to chitoheptaose. As evidenced by the K_m values, the galactosylation improves considerably when going to higher oligomers, which increasingly become better substrates. Along the same line, the relative rates of transfer decrease only to about 70%. In preparative experiments on the mg-scale, final yields of the tetra-, penta-, and hexasaccharides [Galβ1-4 (GlcNAβ-4)$_{2,3}$ and $_4$GlcNAc] were 63%, 81%, and 98%, respectively (H. Streicher and J. Thiem unpublished data).

A novel type with an unprecedented regiodirection of the transfer employing β,1-4-GalT was observed with N-acetyl kanosamine and D-xylose (Fig. 4, upper part; Nishida et al. 1993; Wiemann et al. 1994). In addition to the expected β1-4-galactosylation of xylose, the symmetrical molecule is apparently recognized in a reverse orientation, which renders the anomeric β-OH group structurally similar to the equatorial 4-OH group, and this leads to the β1–β1 inter-

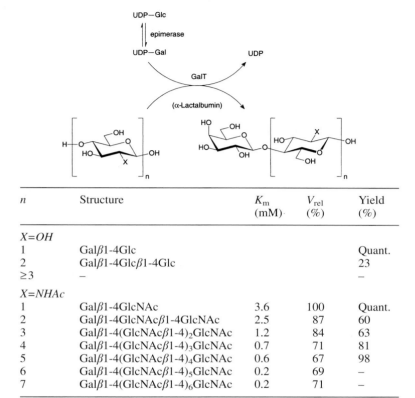

n	Structure	K_m (mM)	V_{rel} (%)	Yield (%)
X=OH				
1	Galβ1-4Glc			Quant.
2	Galβ1-4Glcβ1-4Glc			23
≥3	–			–
X=NHAc				
1	Galβ1-4GlcNAc	3.6	100	Quant.
2	Galβ1-4GlcNAcβ1-4GlcNAc	2.5	87	60
3	Galβ1-4(GlcNAcβ1-4)$_2$GlcNAc	1.2	84	63
4	Galβ1-4(GlcNAcβ1-4)$_3$GlcNAc	0.7	71	81
5	Galβ1-4(GlcNAcβ1-4)$_4$GlcNAc	0.6	67	98
6	Galβ1-4(GlcNAcβ1-4)$_5$GlcNAc	0.2	69	–
7	Galβ1-4(GlcNAcβ1-4)$_6$GlcNAc	0.2	71	–

Fig. 3. Galactosyltransfer with β1-4GalT on cello- and chito-oligomers

glycosidic linkage of the novel, nonreducing disaccharide. Recognition of the acceptor substrate in the normal orientation apparently prevails, thus giving Galβ1-4Xyl:Galβ1-1βXyl in a ratio of about 2:1. More recently these experiments were repeated employing a lactating Holstein cow and here a ratio of about 4:1 was observed (Hara and Suyama 2000).

As with some other completely unexpected substrates, the N-acetylated β-glycosylamines of L-glucose and L-xylose could be shown to undergo β-galactosylation (Fig. 4, lower part). Amazingly,

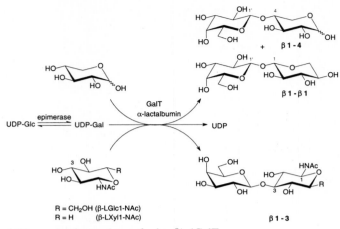

Fig. 4. Unusual galactosyl transfer by β1-4GalT

the recognition of these acceptors by the enzyme occurred to give β1-3-interglycosidically linked disaccharides in about 20% yield (Nishida et al. 2000).

There have been reports from other groups as to the donor specificity concerning UDP-Gal as well as to the acceptor specificity regarding βGlcNAc. These were compiled and extensively discussed (Palcic and Hindsgaul 1996), and Fig. 5, partially based on these data, shows some UDP-Gal modifications to be excellent donors, whereas for others the rates are considerably low, yet sufficient for preparative purposes. It was further shown that the enzyme can recognize and accept a wide range of modified, substituted, and even glycosylated structures beyond the genuine GlcNAc-acceptor substrates for galactosylation.

To summarize, GalT represents an enzyme with an exceptional substrate tolerance, and its synthetic use became obvious early on and continues to be of relevance.

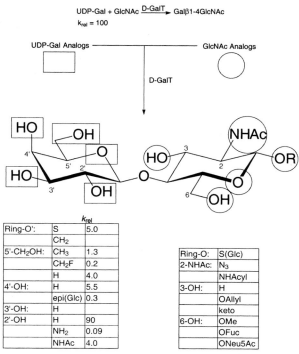

Fig. 5. Donor and acceptor specificities of β1-4GalT

5.3 α2-3-Sialyltransferase

In contrast to the other glycosyltranferases, the four enzymes of this group use a nucleotide monophosphate CMP-Neu5Ac as donor. Whereas α2-3-SiaT (EC 2.4.99.6) and α2-6-SiaT (EC 2.4.99.1) α-sialylate LacNAc structures (Galβ1-4GlcNAc) at positions 3 or 6, respectively, the α2-3-SiaT (EC 2.4.99.4) and α2-6-SiaT (EC 2.4.99.3) affect α-sialylation at positions 3 of O-linked glycoproteins with terminal Galβ1-3GalNAc structures or at position 6 of terminal αGalNAc residues.

En route to the synthesis of sialylated Thomsen-Friedenreich antigen structures, the formation of α-glycosides of serin or threonin of

N-acetyl galactosamine was required. Their previously described enzymatic synthesis from GalNAc and the amino acid reported rather low yields (Johansson et al. 1991). Therefore, a classical eight-step, synthetic route from galactose was selected which resulted in the glycosides in enhanced overall yields. Further transformation employing a β1-3-galactosidase in transglycosylation mode led to the Galβ1-3GalNAcα1-OSer/Thr derivatives in 20%–30% yield (Gambert and Thiem

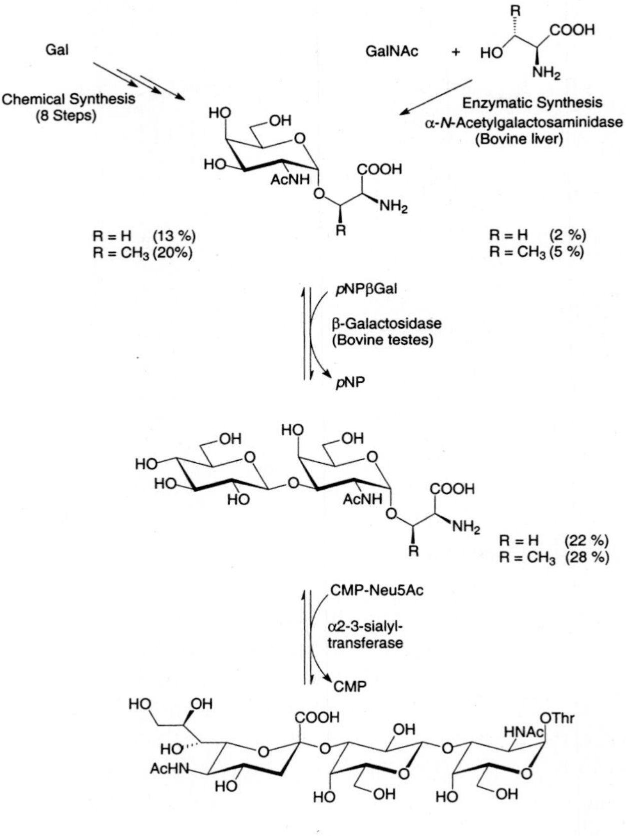

Neu5Acα2-3Galβ1-3GalNAcα1-OThr   (24 %)

Fig. 6. Chemoenzymatic formation of sialylated Thomsen-Friedenreich antigen

1997). Treatment of this acceptor substrate with CMP-Neu5Ac and $\alpha 2$-3-SiaT from porcine liver led to the desired trisaccharide threonine glycoside (Gambert and Thiem 1999; Fig. 6).

Previously, the preparation of the pure trisaccharide Neu5Ac$\alpha 2$-3Galβ1-3 GalNAc from p-nitrophenyl galactoside and N-acetyl-galactosamine could be elaborated with the combined application of transglycosylating glycosidases and selectively transferring glycosyltransferases (Kren and Thiem 1995). Employing six enzymes in one port at optimized pH, this reaction resulted in the desired compound in isolated 45% yield. Extension of this approach to the threonin α-N-acetyl-galactosaminide could be achieved as well and led to the decisive structures of O-linked glycopeptides (Gambert and Thiem 1999; Fig. 7).

This elaborated approach could then be adopted for the formation of glyco-coated bovine serum albumin as neoglycoconjugates. Starting

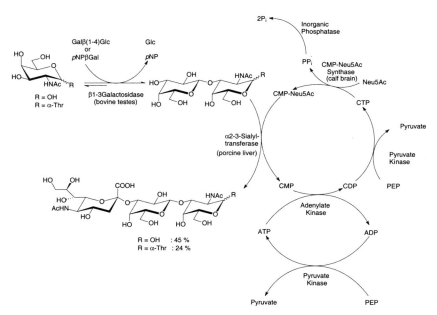

Fig. 7. Cofactor regenerated formation of Neu5Acα2-3Galβ1-4GlcNAc and sialylated Thomsen-Friedenreich antigen

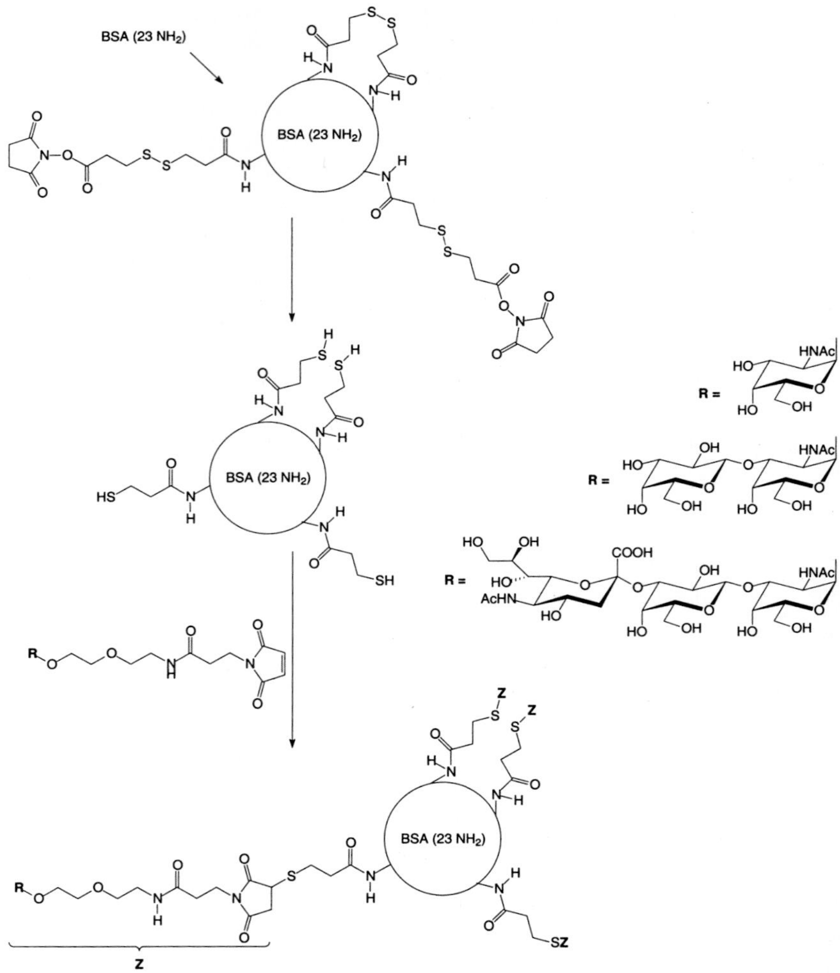

Fig. 8. Mono-, di- and trisaccharide-coated BSA as neoglycoconjugates

from galactose azido-ethylene glycol-spacered glycosides of GalNAc could be prepared and enzymatically galactosylated (Gambert et al. 1997). Further, as above SiaT-mediated sialylation resulted in the corresponding trisaccharide. In all three glycosides, the azido-spacer was reduced to the corresponding terminal amino spacer which by amino-lysis with *N*-succinimidyl-3-maleidopropionate resulted in the maleinimido-spacered saccharides with terminal GalNAc, Galβ1-3GalNAc, and Neu5Acα2-3Galβ1-3GalNAc sugar epitopes.

Simultaneously in BSA (M ~ 67 kDa), the amino groups of lysine were treated with dithiopropionic acid succinimidyl ester to give mono- or difunctional groups. By reductive cleavage of these with dithiothreitol the terminal SH-groups on the functionalized BSA were released. Their coupling by Michael addition to the mono-, di-, or trisaccharide carrying the spacered Michael acceptor function led to the novel saccharide-coated BSA derivatives as neoglycoconjugates to be used for MAB formation and further studies (Gonzales Lio and Thiem 1999; Fig. 8).

Some further discussions of earlier experiments regarding the substrate specificity of the SiaTs were compiled (Palcic and Hindsgaul 1996).

5.4 α1-3/4-Fucosyltransferase

Enzymes of this class (FucT, EC 2.4.1.65), conveniently isolated from human milk, are required to finalize oligosaccharide compounds to construct blood groups-related antigenic determinants such as Lewis-a and Lewis-x. The prerequisite for syntheses with FucT is the donor GDP-Fuc, which is made by nature as depicted in Fig. 9. The de novo synthesis starting from glucose or mannose leads to GDP-Man, which in turn by a three-enzyme complex is transformed into GDP-Fuc.

Studies using this enzyme complex of the green algae *Chlorella vulgaris* for transforming GDP-Man (Barker and Hebda 1982) as depicted resulted in only 4% transfer, which was insufficient for further preparative attempts.

In following the alternative salvage pathway, the required enzymatic synthesis to GDP-Fuc could be achieved from fucose using two enzymes from porcine submaxillary glands. First, fucokinase forms the β-L-fucopyranosyl phosphate and this in turn, with GTP and catalyzed by GDP-fucose-pyrophosphatase, leads to GDP-Fuc. The reaction is further promoted by enzymatic cleavage of the re-

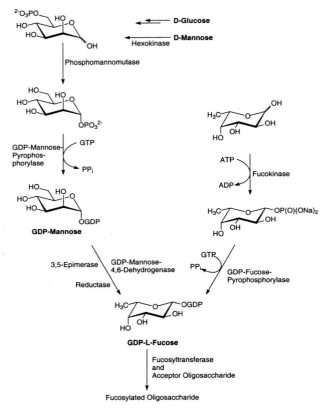

Fig. 9. Natural formation of GDP-L-Fuc

sulting pyrophosphate to inorganic phosphate (Fig. 10; Stiller and Thiem 1992).

Due to the tedious enzymatic reactions which gave yields in the range of 20% and because of interests in modified donor structures, classical construction of GDP-Fuc and derivatives were followed (Fig. 11). Reduction of D-galacturonic acid gave L-galactono-1,4-lactone from which several synthetic steps led to modified L-galactosyl phosphates, and these in turn were transformed into their GDP derivatives employing the morpholidate method (Moffat and Khorana

PYR
ATP
Pyruvate
Kinase
PEP
ADP
Fucokinase
(Porcine Submax.
Glands)
GTP
GDP-
Fucose-Pyro-
phosphorylase
(Porcin Submax.
Glands)
Inorganic
Pyrophosphatase
2 P_i
PP_i

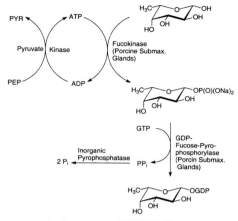

Fig. 10. Chemoenzymatic formation of GDP-L-Fuc

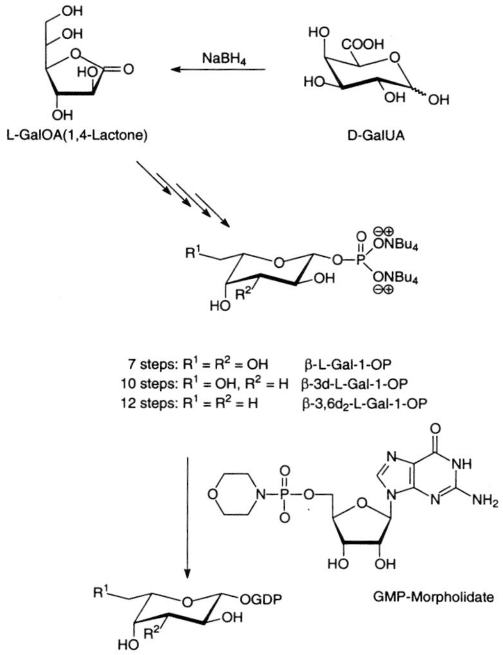

NaBH$_4$
L-GalOA(1,4-Lactone)
D-GalUA
7 steps: R^1 = R^2 = OH β-L-Gal-1-OP
10 steps: R^1 = OH, R^2 = H β-3d-L-Gal-1-OP
12 steps: R^1 = R^2 = H β-3,6d$_2$-L-Gal-1-OP
GMP-Morpholidate

Fig. 11. Chemical synthesis of deoxygenated GDP-L-Gal derivatives

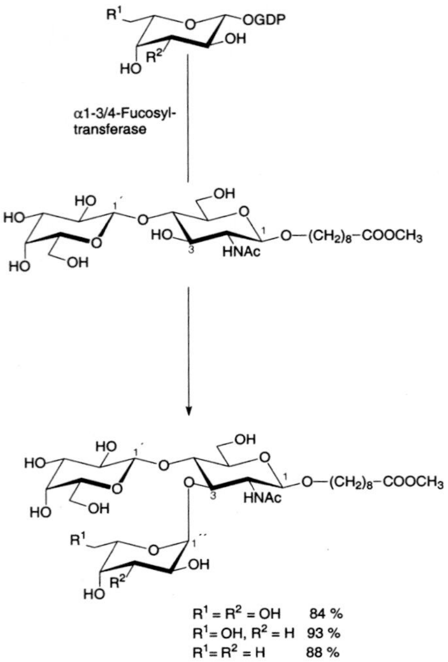

Fig. 12. Enzymatic formation of modified Lewis-x trisaccharide structures

1961). Thus, straightforward pathways to the GDP derivatives of L-galactose, as well as its deoxy and dideoxy analogs could be elaborated (Binch et al. 1999).

Taking as LacNAc acceptor glycoside the β-8-methoxycarbonyloctyl derivative, transfer of the modified fucose derivatives was studied with α1-3/4-fucosyltransferase from human milk (Fig. 12). Each of the donors showed transfer to the 3 position of GlcNAc, and in convincing yields of 84%–93% the modified Lewis-x trisaccharides could be obtained (Stangier et al. 1998).

Employing a recombinant human α1-3-fucosyltransferase (Weston et al. 1992) acceptor substrate studies showed a broad acceptor specificity (Table 1) and this enzyme could be also utilized for mg-scale preparations of Lewis-x oligosaccharides (Wong et al. 1992).

Table 1. Acceptor specificity of human α1-3 FucT

Acceptor substrates	K_m(mM)	V_{rel}(%)
Galβ1-4GlcNAc	35	100
Neu5Acα2-3Galβ1-4GlcNAc	100	620
Neu5Acα-6Galβ1-4GlcNAc	70	13
Galβ1-4GlcNAcβ1-OAll	16	64
Neu5Acα2-3Galβ1-4GlcNAcβ1-OAll	280	380
Galb-3GlcNAc	600	130
Galβ1-4Glc	500	150
Galβ1-4Glc(5S)	12	51
Galβ1-4Glc-al	34	10
Neu5Acα2-3Galβ1-4Glc-al	64	330

With the β-8-methoxycarbonyloctyl glycoside of isolactosamine (Galβ1-3GlcNAc) transfers of GDP-Fuc, GDP-3d-Fuc, and GDP-D-Ara employing α1-3/4 fucosyltransferase (human milk) were performed. The relative rates were 100:2.3:5.9, yet the preparative synthesis of disaccharide glycosides glycosylated at the 3 position could be shown to work well in the mg scale (Gokhale et al. 1990) (Table 1).

5.5 α1-2-L-Galactosyltransferase

In the albumen glands of snails, galactans are produced which are used as the almost exclusive source of nutrition for embryonic and newly hatched snails (Bretting et al. 1985). For instance, the galactan of *Helix pomatia* consists of D-Gal ($\sim 85\%$) and L-Gal ($\sim 15\%$) residues. The D-Gal moieties are β1-3- and β1-6-linked, giving an irregular and highly branched structure, and the L-Gal residues are exclusively found in terminal, nonreducing positions α1-2-glycosidically linked to the prior D-Gal unit.

In earlier studies it was shown that the albumen glands could be directly used for preparative purposes (Bornaghi et al. 1998). Employing GDP-Fuc as modified donor, the decisive structural elements of the human H-blood group determinant could be obtained favorably.

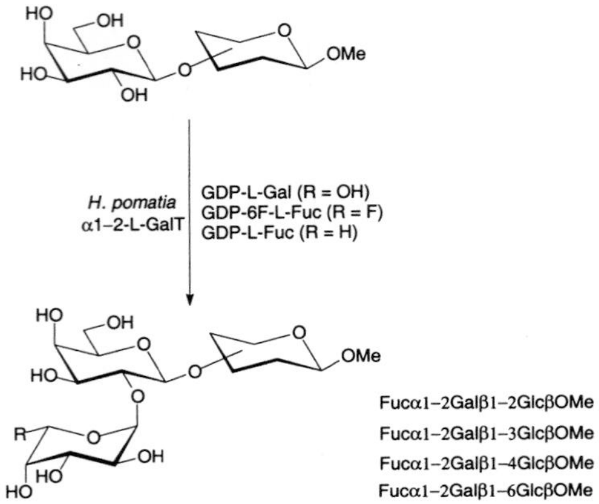

Fig. 13. Fucosylation of Galβ1-nGlcβOMe with α1-2LGalT from *Helix pomatia*

In a previous paper on syntheses with porcine α1-2-fucosyltransferase (α1-2-FucT, EC 2.4.1.69), a trisaccharide formation from Lac-NAc-β-hexanolamine was reported in 83% yield (Rosevear et al. 1982). However, such enzymes from other mammals were never used, and now this snail L-GalT can be employed for fucosylations.

Recent studies revealed that the enzyme also accepted GDP-6-fluoro-L-fucose as donor, which opens the option for further donor modifications. On the acceptor side, a systematic preparation provided the four disaccharide glycosides Galβ1-nGlcβOMe ($n = 2, 3, 4$, and 6). These underwent facile fucosylation at 2′-position regio- and stereoselectively in yields ranging from 65%–71% (Fig. 13). Also Galβ1-4Man could be fucosylated in 67% yield. Acceptor structures, however, which either in the terminal or in the reducing unit show an axial interglycosidic linkage were not recognized.

Tammar wallaby milk contains the galactan-like hexasaccharide (Galβ1-3)$_4$Galβ1-4Glc, which could be D-galactosylated to a decasaccharide and which in turn, by treatment with an excess of GDP-Fuc and α1-2-L-GalT, led to the tridecamer with terminal L-fucose residues (A. M. Scheppokat, H. Bretting, J. Thiem, unpublished data; Fig. 14).

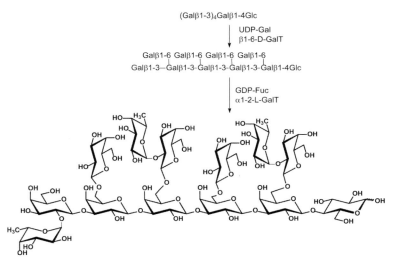

Fig. 14. Subsequent galactosylation and fucosylation with β1-6-D-GalT and with α1-2L-GalT of a galactan – like hexasaccharide

5.6 Conclusion

The cases studied showed a remarkable flexibility of glycosyltransferases both with regard to donor and acceptor substrates. Since many deviations were reported for GalT, the most readily available glycosyltransferase, a considerable number of "erroneous" or "incorrect" transfer reactions may be expected for other enzymes as well. The original and generally accepted concept of "one glycosyl enzyme–one glycosyl linkage" (Natsuka and Lowe 1994) is apparently circumvented more often. This may give rise to speculations concerning the uniform formation and integrity of oligosaccharide conjugates with regard to their biological functions.

From the point of synthesis, the rather low rates of transfer (sometimes around 1% or less) compared with the natural reaction do not prevent the use of glycosyltransferases for the preparation of glycoconjugates or their mimetics in milligram quantities. Scale-up approaches can be expected to result in the formation of larger quantities.

More glycosyltransferases can be expected to be characterized and the most attractive ones cloned for preparative purposes. An approach using glycosyltransferases has been developed and will remain a useful addition and sometimes even a favorable alternative to classical synthesis. Nevertheless, the substrate diversity of glycosyltransferases are still too restricted for access to all desired structural analogs, and thus synthetic approaches will be further required in their own right.

Acknowledgements. Support of our own work by the Deutsche Forschungsgemeinschaft and the Fonds der Chemischen Industrie is gratefully acknowledged.

References

Barber GA, Hebda PA (1982) GDP-D-Mannose: GDP-L-galactose epimerase from Chlorella pyrenoidosa. Methods Enzymol. 83:522–525

Beyer TA, Sadler JE, Rearick JI, Paulson JC, Hill RL (1982) Glycosyltransferases and their use in assessing oligosaccharide structure and structure-function relationships. Adv Enzymol 52:23–175

Binch H, Stangier K, Thiem J (1998) Chemical synthesis of GDP-L-galactose and analogues. Carbohydr Res 306:409–419

Bornaghi L, Keating L, Binch H, Bretting H, Thiem J (1998) Regioselective fucosylation using L-galactosyltransferase from Helix pomatia. Eur J Org Chem 2493–2497

Bretting H, Jacobs G, Bennecke I, König WA, Thiem J (1985) The occurence of L-galactose in snail galactans. Carbohydr Res 139:225–236

Gambert U, Gonzales Lio R, Farkas E, Thiem J, Verez Bencomo V, Liptak A (1997) Galactosylation with β-galactosidase from bovine testes employing modified acceptor substrates. Bioorg Med Chem 5:1285–1291

Gambert U, Thiem J (1997) Chemoenzymatic synthesis of the Thomsen-Friedenreich antigen determinant. Carbohydr Res 299:85–89

Gambert U, Thiem J (1999) Multienzyme system for the synthesis of the sialylated Thomsen-Friedenreich antigen determinant. Eur J Org Chem 107–110

Gokhale UB, Hindsgaul O, Palcic MM (1990) Chemical synthesis of GDP-fucose analogs and their utilization by the Lewis $a(1 \rightarrow 4)$ fucosyltransferase. Can J Chem 68:1063–1071

Gonzales Lio R, Thiem J (1999) Chemoenzymatic synthesis of spacer-linked oligosaccharides for the preparation of neoglycoproteins. Carbohydr Res 317:180–190

Hara Y, Suyama K (2000) Biosynthesis of β1,4- and β1,β1-galactopyranosyl xylopyranosides in the mammary gland of lactating cow. Eur J Biochem 267:830–836

Johansson E, Hedbys L, Larsson PO (1991) Enzymic synthesis of monosaccharide-amino acid conjugates. Enzyme Microb Technol 13:781–787

Kleene R, Berger EG (1993) The molecular and cell biology of glycosyltransferases. Biochim Biophys Acta 1154:283–325

Kren V, Thiem J (1995) A multienzyme system for a one-pot synthesis of sialyl T-antigen. Angew Chem Int Ed Engl 34:893–895

Moffat JG, Khorana HG (1961) Nucleoside polyphospates. X. The synthesis and some reactions of nucleoside 5′-phosphoromorpholidates and related compounds. Improved methods for the preparation of nucleoside 5′-polyphosphates. J Am Chem Soc 83:649–659

Natsuka S, Lowe JB (1994) Enzymes involved in mammalian oligosaccharide biosynthesis. Curr Opin Struct Biol 4:683–691

Nishida Y, Wiemann T, Sinnwell V, Thiem J (1993) A new type of galactosyltransferase reaction: transfer of galactose to the anomeric position of N-acetyl kanosamine. J Am Chem Soc 115:2536–2537

Nishida Y, Tamakoshi H, Kitagawa Y, Kobayashi K, Thiem J (2000) A novel bovine β-1,4-galactosyltransferase reaction to yield β-D-galactopyranosyl-(1-3)-linked disaccharides from L-sugars Angew Chem Int Ed Engl 39:2000–2003

Palcic MM, Hindsgaul O (1996) Glycosyltransferases in the synthesis of oligosaccharide analogs. Trends Glycoscience Glycotechnol 8:37–49

Paulsen H (1990) Synthesis, conformation, and x-ray analysis of saccharide chains of glycoprotein core regions. Angew Chem Int Ed Engl 29:823–839

Rosevear PR, Nunez HA, Barker R (1982) Synthesis and solution conformation of the type 2 blood group oligosaccharide αLFuc(2 → 6)βDGal (1 → 4)βDGlcNAc. Biochemistry 21:1421–1431

Scheppokat AM, Bretting H, Thiem J (2002) unpublished

Stangier K, Palcic MM, Bundle DR, Hindsgaul O, Thiem J (1998) Fucosyltransferase-catalyzed formation of L-galactosylated Lewis structures. Carbohydr Res 305:511–515

Stiller R, Thiem J (1992) Enzymatic synthesis of β-L-fucose-1-phosphate and GDP-fucose. Liebigs Ann Chem 467–471

Streicher H, Thiem J (1997) unpublished

Thiem J, Treder W (1986) Synthesis of the trisaccharide Neu5Ac α(2 → 6)Gal β(1 → 4)GlcNAc with immobilized enzymes. Angew Chem Int Ed Engl 25:1096–1097

Thiem J, Wiemann T (1991) Combined chemoenzymic synthesis of N-glycoprotein synthons. Angew Chem Int Ed Engl 30:1163–1164

Thiem J, Wiemann T (1992) Synthesis of galactose-terminated oligosaccharides by use of galactosyltransferase. Synthesis 141–145

Varki A (1993) Biological roles of oligosaccharides: all of the theories are correct. Glycobiology 3:97–130

Weston BW, Nair RP, Larsen RF, Lowe JB (1992) Isolation of a novel human alpha (1,3)fucosyltransferase gene and molecular comparison to the human Lewis blood group alpha (1,3/1,4)fucosyltransferase gene. Syntenic, homologous, nonallelic genes encoding enzymes with distinct acceptor substrate specificities. J Biol Chem 267:4152–4160

Wiemann T, Nishida Y, Sinnwell V, Thiem J (1994) Xylose: the first ambident acceptor substrate for galactosyltransferase from bovine milk. J Org Chem 59:6744–6747

Wong CH, Haynie SL, Whitesides GM (1982) Enzyme-catalyzed synthesis of N-acetyllactosamine with in situ regeneration of uridine 5'-diphosphate glucose and uridine 5'-diphosphate galactose. J Org Chem 47:5416–5418

Wong CH, Dumas DP, Ichikawa Y, Koseki K, Danishefsky SJ, Weston BW, Lowe JB (1992) Specificity, inhibition, and synthetic utility of a recombinant human α-1,3-fucosyltransferase. J Am Chem Soc 114:7321–7322

Wong CH, Halcomb RL, Ichikawa Y, Kajimoto T (1995) Enzymes in organic synthesis: application to the problems of carbohydrate recognition. Angew Chem Int Ed Engl 34:412–423, 521–546

6 Control of FucT-VII Expression in CD4⁺ T Cells

G. S. Kansas

Selectins are carbohydrate-binding glycoproteins intimately and critically involved in the regulation of leukocyte traffic. The selectins themselves are well characterized, and much is known regarding their distribution, function in leukocyte recruitment and inflammation, cell biology, biophysical basis for cell adhesion, and other aspects; these have been the subject of several reviews (Kansas 1996; Lowe 2002; Alon and Feigelson 2002). More recently, attention has focused on the enzymes (glycosyltransferases) which are responsible for the biosynthesis of the carbohydrate ligands for selectins. Although the complete structure of the carbohydrate ligands for selectins is still not completely understood, much progress has been made in identifying the enzymes responsible for their synthesis. This review focuses principally on one such enzyme, the $\alpha 1,3$ fucosyl-transferase FucT-VII, and control of its expression in T cells.

6.1 Expression of Glycosyltransferases Is Regulated Primarily at the Transcriptional Level

The activity of many classes of enzymes is controlled by various posttranslational modifications, including phosphorylation, sulfation, and glycosylation. In particular, many enzymes involved in transmitting signals from the cell's exterior to the interior and ultimately to the nucleus are activated or deactivated by the addition or removal of one or more phosphate groups at specific residues, and this contributes greatly to the fidelity and specificity of signal transduction.

In contrast, the activity of glycosyltransferases appears to be controlled primarily, if not exclusively, by whether the gene is transcribed or not. There is very little information on the possible role of mRNA stability or translational control of glycosyltransferase genes, and no compelling evidence that these mechanisms, well established for other genes, play any significant role in determining the level of activity of any characterized glycosyltransferase, although it should be emphasized that this has not yet been carefully or extensively examined. Hence, to a first approximation, transcriptional control of expression of a given glycosyltransferase is tantamount to control of enzyme activity.

This should not be taken to mean that no other factors influence the amount of biological product of any given glycosyltransferase which accumulates or is present in a cell. Obviously, if substrate is limiting or if there are mutations in other essential components of the pathway, then the amount of product will be affected. However, in the normal situation, a direct relationship can be assumed to exist between the steady state levels of the mRNA encoding a given glycosyltransferase and the activity of the enzyme. Hence, transcriptional activity is the single greatest determinant of enzyme activity. This conclusion has been documented for FucT-VII (Knibbs et al 1996; Wagers et al 1997), the subject of this review.

6.2 Pattern of Expression of FucT-VII

Although there are at least nine cloned fucosyltransferases, each with it own pattern of expression, the pattern of expression of FucT-VII is consistent with the notion that FucT-VII is specialized for biosynthesis of all classes of selectin ligands on all cells which express selectin ligands. Thus, in both mice and humans, FucT-VII is expressed constitutively in high endothelial venules (HEV), the specialized postcapillary venules present in secondary lymphoid organs such as lymph nodes. In lymph node HEV, FucT-VII makes an essential contribution to ligands for L-selectin (Maly et al 1996; Smith et al 1996). In this setting, FucT-VII supports the normal recirculation of (mostly naive) T and B cells throughout the secondary lymphoid organs, a process essential for effective immunity.

FucT-VII is also expressed constitutively in mature myeloid cells, including neutrophils and monocytes, where it is responsible for maintaining expression of ligands for all three selectins, an essential component of leukocyte recruitment in diverse inflammatory settings. Although it has not been directly examined, FucT-VII is probably also expressed in immature hematopoietic cells extending as far back in differentiation as the hematopoietic stem cell (HSC). This inference is based on the well-documented role of the endothelial selectins in promoting homing of HSC to bone marrow in the setting of bone marrow transplantation (Frennette et al 1998), and the essential role of FucT-VII in construction of all characterized selectin ligands (see Sect. 6.3). Expression of FucT-VII during hematopoiesis is an important question, and its answers have important implications for the control of FucT-VII expression.

In contrast to HEV and myeloid cells, expression of FucT-VII in lymphocytes is inducible and regulated by multiple external signals. Naive T and B cells express little or no FucT-VII, and have correspondingly undetectable levels of selectin ligands (Wagers et al 1998; Lim et al 1999). Activation of T cells induces FucT-VII, and levels of FucT-VII are modulated both up and down by specific cytokines. It is not known whether activation of B cells also induces FucT-VII expression, and little is known about B cell traffic; B cells are rarely seen in sites of acute or chronic inflammation. However, plasma cells, the terminally differentiated descendants of B cells, are often found at site of chronic inflammation such as the rheumatoid synovium in rheumatoid arthritis, and express high levels of FucT-VII, and correspondingly bind well to E-selectin (Underhill et al 2002). The stage of antigen-driven B cell activation and differentiation at which FucT-VII is expressed, and the stimuli which induce expression, are unknown, and represent important questions.

As far as is known, FucT-VII is not expressed in at least most other tissues or cell types, such as fibroblasts, epithelial cells, or neural cells (Sasaki et al 1994; Natsuka et al 1994). This lends further support to the idea, mentioned above, that FucT-VII is specialized for biosynthesis of selectin ligands.

6.3 FucT-VII Is Essential for the Construction of all Known Selectin Ligands

In addition to being specialized for selectin ligand biosynthesis, FucT-VII is essential for the construction of selectin ligands of all classes in all cells in which it has been examined. This includes L-selectin ligands in HEV, ligands for all three selectins expressed on monocytes, neutrophils and other myeloid cells, and E- and P-selectin ligands on activated T and B cells.

This is exemplified most clearly in mice with homozygous null alleles in FucT-VII (Maly et al 1996). Neutrophils from these mice have sharply reduced (>90%) ligands for both E- and P-selectin, with the remaining low level of activity being attributable to a distinct enzyme, FucT-IV. Activated T cells from these mice have no detectable selectin ligands (Smithson et al 2001; Knibbs et al 1998). The HEV found in lymph nodes of these mice do not stain with an L-selectin/IgM fusion protein, and homing of lymphocytes from the blood across HEV into lymph nodes is sharply impaired. In addition, these lymph nodes are quite hypocellular, due to interrupted lymphocyte traffic into lymph nodes from the blood. The HEV phenotype of the FucT-VII-null mice therefore recapitulates that seen in L-selectin null mice (Arbones et al 1994), confirming the receptor/ligand relationship.

Studies using transfected cell lines have shown that enforced expression of FucT-VII in any cell type thus far examined, regardless of lineage or species, leads to expression of functional ligands for E-selectin (Knibbs et al 1996; Wagers et al 1997; Natsuka et al 1994). In contrast, ligands for P-selectin or L-selectin are not formed on all FucT-VII transfectants. For P-selectin, this lack of ligands on all FucT-VII+ cells can often be explained by the absence of PSGL-1, the glycoprotein ligand for P-selectin (Sako et al 1993; Snapp et al 1996). These findings also indicate that FucT-VII substrates representing potential E-selectin ligands are widely if not ubiquitously expressed. This conclusion has important implications for metastasis of malignant cells, since aberrant expression of FucT-VII or functionally similar enzymes could confer E-selectin binding on such transformed cells.

It is important to emphasize in this context that FucT-VII is the only enzyme thus far identified which has been demonstrated to be essential for all known selectin ligands, irrespective of cell type or selectin. There are multiple other enzymes involved in selectin ligand biosynthesis, including O-linked carbohydrate branching enzymes, sialyltransferases and sulfotransferases, and these are each involved in construction of only a subset of selectin ligands, and are similarly not required for all ligands on all cell types. For example, the O-linked branching enzyme core 2 β1,6 glucosaminyltransferase-I (C2GnTNAcT-I) is essential for P-selectin ligand activity but not for most E-selectin ligand activity on leukocytes or for L-selectin ligand activity on HEV (Ellies et al 1998; Kumar et al 1996; Snapp et al 2001; Sperandio et al 2001). Many of these other enzymes also have other documented substrates and/or functions as well. Hence, FucT-VII is unique both in its requirement in construction of all selectin ligands as well as in its specialization for that function.

6.4 FucT-VII as a Possible Target for Skin T Cell Diseases

Although selectins are widely referred to as being essential components of leukocyte recruitment, this is not strictly true for all leukocytes or for all tissues. For example, recruitment of inflammatory cells to internal organs such as the liver and pancreas or the brain is not dependent on selectins (Erdmann et al 2002; Wong et al 1997; Engelhardt et al 1997). Apart from their role in lymphocyte recirculation (L-selectin) and homing of HSC to bone marrow (E- and P-selectin), which may be considered "steady-state" functions, in the context of inflammation, selectins are more accurately thought of as being of crucial importance in rapid recruitment of leukocytes, particularly neutrophils, to sites of likely entry of pathogens: skin, gut (including peritoneum), and lung. This explains, for example, the chronic skin infections and dermatitis characteristic of mice deficient in both endothelial selectins (Bullard et al 1996; Frennette et al 1996): in the absence of the endothelial selectins, neutrophils are unable to efficiently enter sites of bacterial colonization, resulting in failure to clear these otherwise innocuous infections.

Given the crucial importance of selectins in leukocyte recruitment to skin, and the crucial role of FucT-VII in formation of selectin ligands in T cells, T cell FucT-VII becomes an attractive target for treatment of T cell-mediated skin disorders such as psoriasis and atopic dermatitis as well as malignant T cell diseases which exhibit a tropism for skin, such as mycosis fungoides, cutaneous T cell lymphoma, and adult T cell leukemia/lymphoma. However, drugs which directly target FucT-VII would affect other leukocytes as well, potentially leading to general immune suppression or inhibition of inflammation, which historically has represented a major difficulty in selectin antagonist-based approaches to inflammatory disease. Therefore, targeting of FucT-VII in T cells probably requires targeting T cell-specific genetic and transcriptional mechanisms which control FucT-VII expression in these cells, rather than direct targeting of the enzyme itself. This implies that these mechanisms which control inducible expression of FucT-VII in T cells are distinct from and independent of those that maintain constitutive levels of FucT-VII in both myeloid cells and HEV. Understanding how FucT-VII expression is controlled in these different cell types is therefore of high importance for these approaches.

6.5 Induction of FucT-VII in T Cells

In humans, expression of FucT-VII on T cells is quantitatively associated with cell surface expression of epitopes defined by the HECA-452 mAb (Knibbs et al 1996; Wagers et al 1997). Staining with HECA-452, enzyme activity in whole cell lysates, and mRNA levels correspond across a wide range. This mAb thus serves as a reporter for FucT-VII on a per cell basis. This allows for analysis of gene expression of FucT-VII in human T cells by flow cytometry.

By both staining with HECA-452 and RT-PCR, human naive (CD45RA$^+$) CD4$^+$ T cells do not express FucT-VII, and consequently do not express selectin ligands (Wagers et al 1998). A similar finding was made in the mouse (Lim et al 1999). A subset of resting human memory peripheral blood CD4$^+$ cells are HECA-452$^+$, indicating that expression of FucT-VII is associated with T cell activation but that not all activated T cells express FucT-VII. Human and murine CD4$^+$ T cells activated in vitro with either plate bound mAb or antigen express modest levels of FucT-VII, and these levels are substantially enhanced by interleukin (IL)-12. This provides a partial explanation for why T helper 1 (Th1) cells have much higher levels of selectin ligands than do Th2 cells or unpolarized cells. Unexpectedly, IL-12-induced enhancement of FucT-VII expression in the mouse is unaffected by the absence of Stat4, the major signaling pathway by which IL-12 controls gene expression in T lymphocytes (White et al 2001). It remains unclear how IL-12 controls FucT-VII expression in CD4$^+$ T cells.

Two other cytokines have been shown to control or modulate FucT-VII levels in activated CD4$^+$ T cells. The first is IL-4, which drives Th2 development. Using CD4$^+$ T cells activated in vitro by either plate-bound mAb or by antigen, it was observed that IL-4 strongly inhibits the induction of FucT-VII associated with triggering of the T cell receptor (TCR; Wagers et al 1998; Lim et al 1999). Thus, IL4 and IL-12 drive FucT-VII levels in opposite directions. This observation helps explain why Th2 cells have such low or undetectable levels of selectin ligands (Austrup et al 1997): inhibition of induction of FucT-VII by IL-4 would prevent biosynthesis of all such ligands. Although there are undoubtedly other mechanisms by which these cytokines control selectin ligand expression in Th1 and Th2 cells, the opposite effects of these two polarizing cytokines goes a long way towards explaining the distinct adhesive and migratory properties of these two major classes of CD4 cells.

Screening of multiple other cytokines known to be active on CD4 T cells revealed only one other which had any effect on FucT-VII levels: transforming growth factor (TGF)-β1 (Wagers et al 2000).

TGF-β1 was found to potently induce FucT-VII (and C2GnTNAcT-I) and corresponding selectin ligands in naive human CD4 T cells. Further, this induction could be completely blocked by addition of SB203580, a pyrimidazole compound well known to inhibit p38 MAP kinases, but was unaffected by addition of PD98059, an inhibitor of the MEK kinases upstream of the ERK MAP kinases. This suggested that TGF-β1 operated through a p38 MAPK pathway. Moreover, because both the TCR and the IL-12R also activate one or more MAPK pathways, this observation suggested more broadly that induction of FucT-VII in T cells was controlled at least in part by MAPK pathways.

One important caveat in this is that the dose of SB203580 used to block TGF-β1-induced FucT-VII (20 μM) is somewhat higher than is required to block p38 MAPK activation in these cells, and this compound can affect other, non-MAP kinases as well. In addition, it is always difficult to assign a biologic function to a particular pathway solely on the basis of the ability of a pharmacologic inhibitor to block that function or pathway. As detailed below, newer information forces a revision of the interpretation of those data.

In summary, both the TCR and a limited number of cytokine receptors play a role in controlling the level of FucT-VII expressed in an activated CD4$^+$ T cell. How these different cell surface receptors affect the level of FucT-VII transcription and through what pathways is therefore an important question.

6.6 JSB3: A Novel Jurkat Variant Which Serves as a Model System for FucT-VII Regulation

Because of the considerations discussed above, we wanted a system to study the role of MAPK in FucT-VII induction in a system that was robust and amenable to rapid experimental readout. We took advantage of observations of Stoolman and colleagues (Knibbs et al 1996), who showed that a spontaneously arising variant of the human T cell line Jurkat exhibited induction of FucT-VII in response to phobol esters such as PMA. We therefore used this cell line, named JS9-78, to select for a much more highly PMA-responsive cell line by three rounds of sorting the brightest ∼2% of HECA-452+ cells following treatment of the cells with 10 nM PMA for 2 days. We obtained a cell line, designated JSB3, which shows higher fractions of cells expressing FucT-VII for longer duration following brief (2 h) PMA exposure. We have used this cell line as a model system to characterize signal transduction pathways which control FucT-VII expression.

PMA models a subset of signals which emanate from the TCR, by virtue of its ability to activate Ras and protein kinase C. Based on the data above, we evaluated whether MAPK pathways were involved in induction of FucT-VII by PMA and whether Ras was involved. We found that both treatment with PMA and introduction of constitutively active Ras (H-RasV12) via recombinant retroviruses induced FucT-VII in a subset of JSB3 cells, similar to normal CD4 cells activated in vitro (Wagers et al 1998). In addition, FucT-VII induction in response to either PMA or active Ras was blocked partially by PD98059, partially by high dose SB203580, and essentially completely by the combination of the two. The similarities between FucT-VII induction by Ras and FucT-VII induction by PMA, along with the well-documented ability of PMA to activate Ras, strongly suggests that these two stimuli activate the same pathways leading to FucT-VII induction.

A striking feature of FucT-VII induction by Ras was that only the subset of cells with the highest levels of Ras, identified by GFP co-expressed from the recombinant retrovirus, expressed FucT-VII. This indicates that a threshold of Ras signaling is required to trigger FucT-VII transcription, implying that T cells activated through their TCR but with insufficient signal strength will not express FucT-VII and thus will not acquire the ability to migrate to sites of inflammation. Qualitatively different outcomes resulting from differences in signal strength are common in biology and are well established in T cell development in particular. Specifically, Th1 development is associated with strong and sustained TCR signaling, whereas Th2 development is associated with weaker or more transient TCR signaling (Blander et al 2000; Leitenberg and Bottomly 1999). Strong TCR signaling and IL-12 signaling are therefore associated with strong expression of FucT-VII and selectin ligands concomitant with Th1 differentiation, whereas weak TCR signaling and IL-4 signaling are associated with inhibition of FucT-VII and selectin ligand expression during Th2 differentiation.

The serine-threonine kinase Raf is downstream of Ras and upstream of the ERK MAPK pathway. Inhibition of FucT-VII induction by the MEK inhibitor PD98059 implies that Raf is involved is Ras-induced FucT-VII induction. We confirmed this by showing that constitutively active Raf also induced FucT-VII, and that this FucT-VII induction was blocked to baseline levels with PD98059 alone. In addition, similar to Ras-induced FucT-VII induction, expression of FucT-VII was limited to those cells with the highest levels of Raf. These data indicate that Raf represents one of two pathways by which Ras induces FucT-VII in these cells.

The identity of the other pathway by which Ras induces FucT-VII is not known. The defining feature of this pathway is that it is inhibited by high dose (20–50 μM) SB203580. Although this compound was first defined as a p38 MAPK inhibitor, other observations make it unlikely that p38 MAPK are the target of SB203580 in this system. First, a novel compound which is much more specific for p38 MAPK and works at nanomolar concentrations had no effect on Ras-induced FucT-VII induction. Second, constitutively active versions of upstream activators of p38 MAPK failed to induce FucT-VII, and dominant negative p38 MAPK fails to inhibit. As this second, SB203580-inhibitable pathway is likely the relevant pathway for at least TGF-β1, if not for IL-12 and the TCR as well, identifying it remains an important question.

6.7 Conclusions

FucT-VII is specialized for, and essential for, biosynthesis of all defined selectin ligands. FucT-VII is constitutively expressed in myeloid cells and HEV, but inducibly expressed and regulated in lymphocytes. In CD4$^+$ T cells, signals through both the TCR and a limited set of cytokine receptors determine the level of FucT-VII expressed in the cell, and this leads to heterogeneity in the outcome with respect to the level of FucT-VII expressed on different cells within the population. Studies to date suggest that MAP kinase pathways, activated both by the TCR and perhaps by at least some cytokine receptors, are responsible at least in part for induction of FucT-VII in these cells. Another pathway, downstream of Ras and other signals, is also involved, but the molecular identity of this pathway is unknown. Unraveling the details of how T cell expression of FucT-VII is controlled may allow for intervention in an array of T cell-mediated diseases characterized by inflammatory destruction of target tissues for which selectins are important mediators of leukocyte recruitment.

References

Alon R, and Feigelson S (2002) From rolling to arrest on blood vessels: leukocyte tap dancing on endothelial integrin ligands and chemokines at sub-second contacts. Semin Immunol 14:93–104

Arbones ML, Ord DC, et al (1994) Lymphocyte homing and leukocyte rolling and migration are impaired in L-selectin (CD62L) deficient mice. Immunity 1:247–260

Austrup F, Vestweber D, et al (1997) P- and E-selectin mediate recruitment of T helper 1 but not T helper 2 cells into inflamed tissues. Nature 385:81–83

Blander JM, Sant'Angelo DB, et al (2000) Alteration at a single amino acid residue in the T cell receptor α complementarity determining region 2 changes the differentiation of naive CD4 T cells to antigen from T helper 1 (Th1) to Th2. J Exp Med 191:2065–2071

Bullard DC, Kunkel EJ, et al (1996) Infectious susceptibility and severe deficiency of leukocyte rolling and recruitment in E-selectin and P-selectin double mutant mice. J Exp Med 183:2329–2337

Ellies LG, Tsuboi S, et al (1998) Core 2 oligosaccharide biosynthesis distinguishes between selectin ligands essential for leukocyte homing and inflammation. Immunity 9:881–890

Engelhardt B, Vestweber D, et al (1997) E- and P-selectin are not involved in the recruitment of inflammatory cells across the blood-brain barrier in experimental autoimmune encephalomyelitis. Blood 90:4459–72

Erdmann I, Scheidegger EP, et al (2002) Fucosyltransferase VII-deficient mice with defective E- P- and L-selectin ligands show impaired CD4+ and CD8+ T cell migration into the skin, but normal extravasation into visceral organs. J Immunol 168:2139–2146

Frenette PS, Mayadas TN, et al (1996) Susceptibility to infection and altered hematopoiesis in mice deficient in both P- and E-selectins. Cell 84:563–574

Frenette PS, Subbarao S, et al (1998) Endothelial selectins and vascular adhesion molecule-1 promote hematopoietic progenitor homing to bone marrow. Proc Natl Acad Sci USA 95:14423–4428

Homeister JW, Thall AD, et al (2001) The alpha(1,3)fucosyltransferases FucT-IV and FucT-VII exert collaborative control over selectin-dependent leukocyte recruitment and lymphocyte homing. Immunity 15:115–126

Kansas GS (1996) Selectins and their ligands: current concepts and controversies. Blood 88:3259–3286

Knibbs RN, Craig RA, et al (1998) α1,3-fucosyltransferase VII dependent synthesis of P- and E-selectin ligands on cultured T lymphoblasts. J Immunol 161:6305–6315

Knibbs RN, Craig RA, et al (1996) The fucosyltransferase FucT-VII regulates E-selectin ligand synthesis in human T cells. J Cell Biol 133:911–920

Kumar R, Camphausen RT, et al (1996) Core 2 β-1,6-N-Acetylglucosaminyltransferase enzyme activity is critical for P-selectin glycoprotein ligand-1 binding to P-selectin. Blood 88:3872–3879

Leitenberg D and Bottomly K (1999) Regulation of naive T cell differentiation by varying the potency of TCR signal transduction. Semin Immunol 11:283–292

Lim Y-C, Henault L, et al (1999) Expression of functional selectin ligands on Th cells is differentially regulated by IL-12 and IL-4. J Immunol 162:3193–3201

Lowe Jb (2002) Glycosylation in the control of selectin counter-receptor structure and function. Immunol Rev 186:19–36

Maly P, Thall AD, et al (1996) The α(1,3) fucosyltransferase FucT-VII controls leukocyte trafficking through an essential role in L-, E-, and P-selectin ligand biosynthesis. Cell 86:643–653

Natsuka S, Gersten KM, et al (1994) Molecular cloning of a cDNA encoding a novel human leukocyte alpha-1,3-fucosyltransferase capable of synthesizing the sialyl Lewis x determinant. J Biol Chem 269:16789–94

Sako D, Chang X-J, et al (1993) Expression cloning of a functional glycoprotein ligand for P-selectin. Cell 75:1179–1186

Sasaki S, Kurata K, et al (1994) Expression cloning of a novel alpha 1,3-fucosyltransferase that is involved in biosynthesis of the sialyl Lewis x carbohydrate determinants in leukocytes. J Biol Chem 269:14730–14737

Smith PL, Gersten KM, et al (1996) Expression of the α1,3 fucosyltransferase Fuc-TVII in lymphoid aggregate high endothelial venules correlates with expression of L-selectin ligands. J Biol Chem 271:8250–8255

Smithson G, Rogers CE, et al (2001) FucTVII is required for T helper 1 and T cytotoxic 1 lymphocyte selectin ligand expression and recruitment in inflammation, and together with FucT-IV, regulates naive T cell trafficking to lymph nodes. J Exp Med 194:601–614

Snapp KR, Heitzig CE, et al (2001) Differential requirements for the O-linked branching enzyme core 2 β1-6-N-glucosaminyltransferase

(C2GlcNAcT-I) in biosynthesis of ligands for E-selectin and P-selectin. Blood 97:3806–3812

Snapp KR, Wagers AJ, et al (1996) P-selectin glycoprotein ligand-1 (PSGL-1) is essential for adhesion to P-selectin but not E-selectin in stably transfected hematopoietic cell lines. Blood 89:896–901

Sperandio M, Thatte A, et al (2001) Severe impairment of leukocyte rolling in venules of core 2 glucosaminyltransferase-deficient mice. Blood 97:3812–3819

Underhill GH, Minges Wols HA, et al (2002) IgG plasma cells display a unique spectrum of leukocyte adhesion and homing molecules. Blood 99:2905–2912

Wagers AJ and Kansas GS (2000) Potent induction of FucT-VII by TGF-β1 through a p38 MAP kinase-dependent pathway. J Immunol 165:5011–5016

Wagers AJ, Stoolman LM, et al (1997) Expression of leukocyte fucosyltransferases regulates binding to E-selectin. Relationship to previously implicated carbohydrate epitopes. J Immunol 159:1917–1929

Wagers AJ, Waters CM, et al (1998) IL-12 and IL-4 control T cell adhesion to endothelial selectins through opposite effects on FucT-VII gene expression. J Exp Med 188:2225–2231

White SJ, Underhill GH, et al (2001) Cutting edge: differential requirements for Stat4 in expression of glycosyltransferases responsible for selectin ligand formation in Th1 cells. J Immunol 167:628–631

Wong J, Johnston B, et al (1997) A minimal role for selectins in the recruitment of leukocytes into the inflamed liver microvasculature. J Clin Invest 99:2782–2790

7 Selectin Avidity Modulation by Chemokines at Subsecond Endothelial Contacts: A Novel Regulatory Level of Leukocyte Trafficking

O. Dwir, V. Grabovsky, R. Alon

7.1 Introduction

Immune cells (leukocytes) and hematopoietic progenitor cells circulating the body must exit blood vessels near specific target sites of injury, infection, inflammation, or proliferation (Springer 1994; Butcher and Picker 1996; Mazo and von Andrian 1999). Recruitment of different subsets of leukocytes and circulating malignant cells to these sites is tightly regulated by sequential adhesive interactions between specific protein receptors on their surface and respective ligands on the blood vessel endothelial wall (Muller et al. 2001). Accumulated data from in vivo and in vitro studies suggest that the major players in this multistep process are members of two adhesive families, selectins and integrins, which are structurally and functionally adapted to operate under disruptive shear forces exerted on leukocytes at the vessel wall by the blood flow. The primary attachment or tethering of circulating leukocytes to the vessel wall is labile, mediated by specialized adhesive lectins, selectins, permitting leukocytes to roll in the direction of flow and bringing them into proximity with activating chemoattractants or chemokines on the endothelial surface (Mackay 2001). These vessel wall-displayed cytokines bind specific G-protein coupled receptors (GPCRs) on recruited leukocytes and trigger, within subseconds, the activation on the leukocyte surface of a second class of adhesion receptors, integrins, which can then firmly bind to their endothelial ligands causing the immune cell to temporarily arrest on the blood vessel (Campbell and Butcher 2000). A remarkable feature of these receptors is that their activity is dynamically regulated independent of their level of surface expression (Shimizu et al. 1999). This allows immune cells to rapidly adapt their adhesive behavior towards specific endothelial sites within target tissues accordingly to tissue- and context-restricted patterns of chemokine or chemoattractant expression at these sites.

7.2 Specialization of Selectins in Tethering Cells from The Flow to Adhesive Endothelial Surfaces Under High Shear Flow

Selectins are the major adhesive receptors in the vasculature that mediate the initial tethering of flowing leukocytes to the vessel wall and the propagation of these tethers into continuous rolling adhesions (Kansas 1996). Remarkably, the majority of selectin interactions specialized to capture (tether) circulating cells from the bloodstream bind their respective ligands on endothelial or leukocyte scaffold proteins with extremely low affinity (Nicholson et al. 1998; van der Merwe 1999). The high efficiency by which selectins interact under shear flow with their counter-receptors has been attributed to biophysical properties such as high tolerance of their tether bonds to applied shear forces (Alon et al. 1995 b) as well as to cellular properties, in particular, presentation on leukocyte microvilli (von Andrian et al. 1995). These surface projections not only facilitate leukocyte collision and encounter of endothelial ligands but may also function as shock absorbers, reducing the effective shear forces applied on the selectin or its ligand (Shao et al. 1998). Since the duration of initial contact between selectin and its ligand falls in the range of microseconds (Hammer and Lauffenburger 1989), selectin bonds should have a high intrinsic association rate. Paradoxically, however, rates of L- and E-selectin association with soluble ligands are extremely low (in the range of 10^5 s^{-1} M^{-1}; Nicholson et al. 1998; Wild et al. 2001), suggesting that other specialized biophysical and/or cellular properties of selectins, not detected by conventional biochemical assays, must account for their extremely efficient ability to tether cells from the flow.

7.3 Kinetic Measurements of Selectin Tethers Under Shear Flow: A Tool to Dissect the Molecular and Cellular Basis of Selectin Adhesiveness

The binding of selectins to their carbohydrate ligands are the fastest cell–cell recognition events known in nature. The multicomponent nature of leukocyte interactions with target endothelial sites in vivo does not allow dissection of the biophysical properties accounting for selectin function under shear flow. Using flow chamber assays simulating blood flow at sites of leukocyte emigration (Lawrence and Springer 1991) and high resolution videomicroscopic analysis of cellular motions over surfaces reconstituted with known densities of purified endothelial-derived selectins or selectin ligands, several labs have gained new key insights into the molecular basis of selectin adhesiveness (Alon et al. 1995 a, b, 1997, 1998; Chen and Springer 1999; Chen and Springer 2001; Dwir et al. 2000; Dwir et al. 2001, 2002; Greenberg et al. 2000; Lawrence and Springer 1991, 1993; Lawrence et al. 1997; Patel et al. 1995; Puri and Springer 1996; Puri et al. 1997, 1998; Ramachandran et al. 1999, 2001; Setiadi et al. 1998; Yago et al. 2002). A powerful approach to elucidate the inherent properties of selectin-mediated interactions under shear flow has been to measure the kinetics of individual reversible tethers mediated by selectin ligand interactions at subsecond adhesive contacts (Alon et al. 1995 b; Kaplanski et al. 1993; Fig. 1). These transient tethers are the smallest unit of adhesive interaction observable in shear flow. Site density measurements of selectins or ligands indicate that these tethers are mediated by singular adhesive pairs with durations ranging from 10^1 to 10^3 ms (Alon et al. 1995 b; Dwir et al. 2001). Our studies on L-selectin interactions also indicated that these tethers share intrinsic kinetic properties which closely correlate with dynamic properties of rolling adhesions on physiological densities of L-selectin or L-selectin ligands (Alon et al. 1997, 1998; Chen and Springer 2001). Advanced measurements of transient tether kinetics in flow chamber assays at high temporal resolution using high-speed videomicroscopy has been recently introduced (Smith et al. 1999). Millisecond resolution of cell motions over low-density ligands has allowed the detection of millisecond-long tethers mediated by mutated selectins or ligands not resolved in standard videomicroscopy (20–30 ms resolution). These new approaches can provide complementary information to that gained by standard videomicroscopy, in particular on specific structural motifs of selectins or ligands necessary for the conversion of ligand recognition (10^0 ms range) into functional adhesive tethers (10^2 ms range; Dwir et al. 2001, 2002).

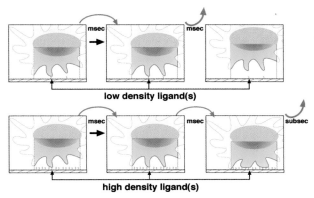

low density ligand(s)

high density ligand(s)

Fig. 1. The transient tether concept: a powerful tool to dissect the initiation and stabilization of selectin-mediated adhesive contacts, the building units of primary leukocyte adhesion to endothelial surfaces under disruptive shear flow. Ligand of interest is irreversibly immobilized on a flow chamber planar substrate. *Top panel*: at low (sub-physiological) densities, rare uniform transient adhesive interactions [quantal adhesive units (Alon et al. 1995 b)] are generated between leukocyte microvilli-based receptor(s) to the adhesive ligand(s). A quantal tether may consist of 1–3 productive microvillar contacts, each containing one or two adhesive bonds (Chen and Springer 1999). Each microvillar site is locally and independently stabilized during the short-lived contact. In addition to receptor density and accessibility on the leukocyte surface, factors such as receptor affinity, anchorage to cytoskeleton, preclustering state, and stretching or bending of microvilli (Shao et al. 1998) will each contribute to these local stabilization events. Tether frequency and duration analysis enable sensitive measurements of local avidity generation at the leukocyte contact with the substrate-bound ligand(s). *Bottom panel*: increasingly higher (physiological) densities of the ligand (or ligands) support engagement of multiple sites, allowing microvilli to deform, and engage additional receptors with the adhesive contact. Thus, hierarchies of local stabilization processes on different ligand densities can be readily dissected by following tether duration at high temporal resolution. (For further details, please refer to Dwir et al. 2001)

7.4 Structural Elements of L-Selectin Regulate Adhesiveness Independent of the Selectin's Affinity to Soluble Ligand

Cumulative findings on L- and P-selectin suggest key regulatory roles for two domains within the selectin ectodomain, the epidermal growth factor (EGF)-like domain and the SCRs, in enhancing the efficiency by which the ligand recognition lectin domains of the selectins at the N' terminus of these adhesion receptors recognize surface-bound ligands at short-lived contacts (Dwir et al. 2000; Patel et al. 1995). Notably, these regulatory domains can alter selectin adhesiveness without changing the apparent affinity of the selectin to soluble ligand, which is controlled by the lectin domain (Kansas et al. 1994). The effective cellular k_{on} of adhesive selectin-mediated tethers to low-density ligand may be enhanced by extending the selectin or its ligand on the interacting adhesive surfaces a sufficient length above the bulk glycocalyx of the plasma membrane (Yago et al. 2002). In addition, specialized mechanical properties of the lectin domain, possibly controlled by its associations with the EGF domain, modulate this effective k_{on} (Dwir et al. 2000). Based on recent crystallographic data, EGF-lectin domain interactions are predicted to enhance the lectin domain flexibility and thereby facilitate selectin binding to immobilized ligand much more than soluble ligand (Somers et al. 2000). This domain flexibility does not modify selectin affinity to soluble ligand, since at a soluble state (three-dimensional), the ligand is much more readily available for selectin binding than at a surface-bound (two-dimensional) state (Bell 1978). Consequently, the ectodomain of selectins as well as their cellular topography may control the effective two-dimensional affinity (2D affinity) of the lectin domain to its surface-bound counter-ligand without changing the apparent 3D affinity, measured by equilibrium binding assays on cell-based or purified selectins (Mehta et al. 1998; Moore et al. 1994; Nicholson et al. 1998; Wild et al. 2001).

7.5 Cytoskeletal Associations of L-Selectin and Multimerization of Its Bonds Regulate Adhesiveness Under Shear Flow

In addition to these considerations, it has become increasingly evident that associations of selectins and ligands with the cell cytoskeleton or with membranal complexes associated with it control selectin adhesiveness under shear flow without altering selectin affinity to soluble ligand (Dwir et al. 2001; Kansas et al. 1993; Setiadi et al. 1998; Snapp et al. 2002). A specific and apparently preformed L-selectin association with the actin cytoskeleton through its cytoplasmic domain (Pavalko et al. 1995) was recently found by us to increase selectin tether duration under flow and to enhance the mechanical stability of the tethers, i.e., their resistance to rapid detachment under increasing shear forces (Dwir et al. 2001). Physical anchorage of a cell-free tail-truncated L-selectin to a solid surface was found both necessary and sufficient to rescue its defective adhesion, suggesting that L-selectin anchorage to the cytoskeleton is obligatory for the selectin tethers to acquire high stability under shear flow. A millisecond stabilization step of individual L-selectin bonds appears essential for leukocyte L-selectin to capture cells and mediate their rolling on different kinds of L-selectin ligands. Notably, the cytoskeletal anchorage of L-selectin does not appear to increase selectin clustering or presentation on tips of microvilli (Pavalko et al. 1995) and does not protect L-selectin from shedding (Dwir et al. 2001). L-selectin shedding on its own does not destabilize L-selectin tethers since protecting L-selectin from proteolytic shedding in various lymphoid cell systems does not stabilize L-selectin-mediated capture and rolling adhesions (Grabovsky et al. 2002) as reported for neutrophils (Hafezi-Moghadam and Ley 1999; Walcheck et al. 1996). The enhanced stability of tethers mediated by cytoskeletally anchored L-selectin is also unlikely to arise from the selectin protection from membrane uprooting, a process much slower than cytoskeletally mediated tether stabilization (Shao and Hochmuth 1999). Stabilization of L-selectin tethers under flow is highly susceptible to mild cell treatment with F-actin disrupting reagents (Dwir et al. 2001; Finger et al. 1996a; Kansas et al. 1993), although E- and P-selectin-mediated rolling is in fact strengthened under particular settings in cells treated with these reagents (Finger et al. 1996a; Yago et al. 2002).

Another means by which selectins and ligands regulate adhesiveness under shear flow is through bond multimerization, which is predicted to disperse highly disruptive shear forces over multiple bonds (Setiadi et al. 1998; Snapp et al. 1998). All known physiological selectin ligands consist of clustered carbohydrate ligands, and selectins are also often found clustered on cell surfaces (Picker et al. 1991; Ramachandran et al. 2001; Setiadi et al. 1998; von Andrian et al. 1995). We have recently found that dimerization of L-selectin enhances tether formation rate to properly spaced ligands without noticeable alteration of the tether duration (Dwir et al. 2002). In contrast, tether duration but not formation rate has been reported to be enhanced by P-selectin or P-selectin ligand dimerization (Ramachandran et al. 2001). Our unpublished results also suggest that alterations in spatial distribution of monovalent PSGL-1-derived glycopeptides can dramatically affect both tether formation rates and tether duration under shear flow. Multimerization of L-selectin ligand can also partially rescue defective tether stabilization caused by selectin tail-truncation (Dwir et al. 2001). However, dimerization of tail-truncated L-selectin fails to enhance tether formation (Dwir et al. 2002). These findings collectively suggest that local generation of selectin avidity at microvillar contacts is critically dependent on proper selectin anchorage to the cytoskeleton as well as on proper spacing of selectin ligands. Thus, in addition to specialized mechanical properties of selectin ectodomains, close spacing of multiple carbohydrate ligands on extended mucin structures (Lasky 1991; Yago et al. 2002), as well as on multiantennary structures (Yeh et al. 2001), may all augment selectin avidity and thereby enhance tether formation and stabilization under shear flow. Cytoskeletal anchorage and clustering states of both L-selectin and its glycoprotein ligands could vary between different leukocytes and endothelial sites, as well as inflammatory conditions (Evans et al. 1999; Li et al. 1998), introducing novel and nonredundant levels of L-selectin adhesion regulation in pathological processes.

7.6 Two-Dimensional Chemokine Signaling to Integrins at Subsecond Contacts

Our recent studies on integrin activation by chemokines in T lymphocytes and hematopoietic progenitors reveal that in situ activation of all major vascular integrins on these cells, i.e., VLA-4, LFA-1 and $\alpha 4\beta 7$, has a clear preference for surface-bound over soluble chemokines (Grabovsky et al. 2000; Peled et al. 1999). Notably, immobilized chemokines, although not adhesive on their own, can increase within subseconds of contact integrin avidity to surface-bound ligand by an in situ Gi signaling event through their GPCR (Campbell et al. 1998; Grabovsky et al. 2000). Locally stimulated avidity of the integrin VLA-4 was found to enhance both leukocyte capture and reversible rolling under shear flow, even in the absence of selectin interactions (Grabovsky et al. 2000). This chemokine potentiation of VLA-4 avidity occurred within less than 0.1 s of contact and involved changes in integrin clustering rather than induction of high affinity to soluble ligand (Grabovsky et al. 2000). Our studies suggest that this subsecond chemokine signaling to integrin requires GPCR proximity to the integrin at the adhesive contact. We therefore refer to this mode of chemokine signaling as 2D-GPCR signaling, since the rapid integrin stimulatory GPCR signals are confined to a 2D interface and cannot be reconstituted with soluble chemokine. We recently found evidence that chemokine signaling to $\alpha 4$ integrins also involves chemokine-induced clustering of GPCRs (Shamri et al. 2002). This clustering is highly sensitive to mild disruption of F-actin (Shamri et al. 2002). Soluble chemokines are incapable of delivering rapid avidity changes to integrins due to inability to induce this GPCR clustering (Shamri et al. 2002). The nature of this event is still unclear, but in an analogy to L-selectin, the 2D-chemokine signal may rearrange the cortical cytoskeleton nearby the target integrin and thereby enhance local integrin clustering at sites of integrin occupancy by ligand. Indeed, both pharmacological and genetic disruption of VLA-4 anchorage to the cytoskeleton strongly interfere with the ability of VLA-4 to tether cells under shear flow (Chen et al. 1999, and Alon et al. manuscript in preparation). The notion that VLA-4 and L-selectin are coexpressed on surface microvilli leads us to hypothesize that chemokine-induced GPCR clustering events on tethered leukocytes may propagate signals to proximal VLA-4 but also to L-selectin at subsecond adhesive contacts.

7.7 Endothelial Chemokines Destabilize L-Selectin-Mediated Leukocyte Rolling Through a Gi-Protein-Independent Nonshedding Mechanism

The multistep paradigm of leukocyte/endothelial adhesion predicts that chemokine activation of leukocyte G-proteins is facilitated by selectin-mediated capture and rolling and does not modulate these primary adhesive steps (Hafezi-Moghadam et al. 2001; Warnock et al. 1998). Our recent results extend this classical paradigm on the role of chemokines, revealing that in addition to stimulating integrin-mediated leukocyte adhesion, surface-bound, but not soluble chemokines, can destabilize the earliest steps of leukocyte rolling mediated by L-selectin on a variety of L-selectin ligands (Fig. 2; Grabovsky et al. 2002). This phenomenon was observed with multiple chemokines and L-selectin expressing leukocytes, including human peripheral blood T lymphocytes, neutrophils, murine T and B splenocytes. and a murine pre-B cell line (Grabovsky et al. 2002). Notably, chemokine suppression of leukocyte rolling did not involve proteolytic L-selectin shedding, since it could not be rescued by blockage of the L-selectin shedding machinery. Rather, it involved the binding of the N' terminus domain of the suppressive chemokines, most probably to its cognate G-protein coupled chemokine receptors (GPCRs; Fig. 3). Unexpectedly, while apparently transduced through dose-dependent binding to specific GPCRs on the responsive leukocyte subset (Grabovsky et al. 2002), the chemokine-suppressive signals did not involve Gi-protein signaling. Chemokine destabilization of L-selectin rolling took place even under complete inhibition of the Gi-signaling machinery on the responsive leukocytes (Grabovsky et al. 2002). Blockage of alternative GPCR signaling to JAK/STAT pathways (Vila-Coro et al. 1999) could also not interfere with chemokine-induced suppression of L-selectin rolling, although mild inhibition of metabolic energy rescued cells from suppression (Grabovsky et al. 2002). Furthermore, a chemokine derivative with no apparent Gi-signaling activity (Grabovsky et al. 2002) and no de-

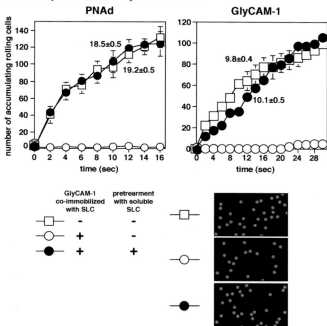

Fig. 2. Immobilized chemokines destabilize L-selectin-mediated PBL rolling on an endothelial L-selectin ligand. Accumulation of rolling human PBL (T lymphocyte) as a function of perfusion time on substrates containing the L-selectin ligands PNAd (coated at 100 ng/ml; left panel) or GlyCAM-1 (100 sites/μm^2; *middle panel*), each coimmobilized with either active (+) or inactivated SLC (–). For CCR7 pretreatment, lymphocytes were preincubated with the CCR7 ligand, SLC, (0.5 μg/ml) in binding medium. Data points represent the means±range of flow experiments performed at a shear stress of 1.75 dyn/cm^2. Mean rolling velocities±SEM are indicated near respective accumulation plots. Representative segments of video frames taken at t=5 s, which depict lymphocytes interacting with GlyCAM-1 under the three experimental conditions mentioned above, are shown at the *right panels*. *Blue images* depict continuously rolling lymphocytes; *green images* depict transiently tethered lymphocytes. (From Grabovsky et al. 2002, with permission)

tectable capacity to transduce GPCR internalization (a phosphorylation-dependent Gi-protein independent process; Brzostowski and Kimmel 2001; Thelen 2001) could also strongly destabilize L-selectin rolling. Chemokine destabilization of L-selectin rolling either re-

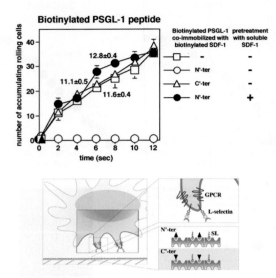

Fig. 3. The GPCR binding domain of SDF-1α mediates its suppressive activity on L-selectin-mediated rolling. Rolling of PBL on a PSGL-1 derived biotinylated peptide (0.1 µg/ml) coimmobilized with SDF-1α biotinylated either in its N′ or C′ terminus. The PSGL-1 peptide was labeled with a single biotin at its C′ terminus (*SL*) and was coimmobilized together with the biotinylated SDF-1α to substrates precoated with avidin. *N′-ter*, SDF-1α, modified in its C′ with biotin, anchored on avidin, leaving its N′ CXCR4 binding domain functionally exposed. *C′-ter*, SDF-1α, modified in its N′ with biotin, and anchored on avidin leaving its C′ terminus functionally exposed. Where indicated, PBL were pretreated with 0.5 µg/ml of soluble SDF-1α. Experiments were all conducted at a shear stress of 1.75 dyn/cm^2. *Inset*: a scheme depicting a cell tethered to a substrate coated with avidin presenting the PSGL-1 peptide and the SDF-1α derivatives. SDF-1α bound via its biotinylated C′ to the substrate and presenting its N′ CXCR4 binding domain (*N′-ter*) is shown as an *upside-up triangle*. SDF-1α presenting its C′ terminus (*C′-ter*) is shown as an *upside-down triangle*. (Based on data from Grabovsky et al. 2002)

sulted in immediate detachment of rolling cells or in faster and jerkier rolling but did not require the presence of integrin ligands, suggesting it resulted from an intrinsic interference with L-selectin adhesiveness. Accordingly, chemokine interference with rolling was not the result of masking carbohydrate moieties on L-selectin li-

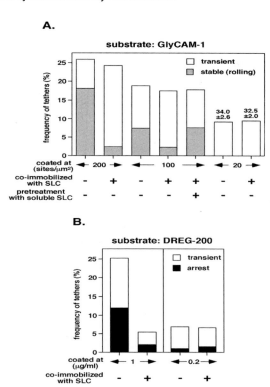

Fig. 4A, B. Chemokine suppresses L-selectin-mediated rolling and adhesion to high-density selectin ligand (or mAb) but does not interfere with transient tethering to low-density ligands. **A** Frequency and type of tethers formed by PBL perfused over substrates coated with the indicated densities of Gly-CAM-1 coimmobilized with SLC or inactivated SLC (−) at a fixed density. Tethers were classified as stable (i.e., followed by rolling) or transient. Mean duration (ms) of transient tethers to GlyCAM-1 (20 sites/μm^2) is indicated on *top* of the *bars*. (Based on data from Grabovsky et al. 2002). **B** Endothelial-displayed chemokine suppresses lymphocyte adhesion to anti-L-selectin lectin domain mAb coated at high density but does not interfere with adhesion to low-density mAb. Tethering frequency (transient or followed by immediate arrest) of PBL perfused over substrates coated with the L-selectin mAb, DREG-200 (Dwir et al. 2000). The mAb was coated at either 1 or 0.2 μg/ml followed by coating with a fixed amount (4 μg/ml) of either active (+) or inactivated SLC (−). Similar results were obtained when SLC was replaced by SDF-1α

gands, since lymphocytes expressing L-selectin but lacking functional GPCR could normally roll on selectin ligands coimmobilized with chemokines (Grabovsky et al. 2002). Suppression of rolling was also more robust when L-selectin ligand density was high, excluding a role for chemokines in directly blocking L-selectin ligand interactions or suppressing inherent 2D L-selectin affinity to ligand (Fig. 4 A). Collectively these data suggest that GPCR occupancy by immobilized chemokines attenuate in situ L-selectin adhesion to high-density but not to low-density endothelial ligands. Attenuation of rolling is therefore predicted to be more robust at endothelial sites expressing high densities of L-selectin ligand, most probably lymph node high endothelial venules (HEV; Stein et al. 2000). Chemokine destabilization of L-selectin adhesions may serve as a negative feedback means to render L-selectin rolling more labile. Faster L-selectin rolling has been recently shown to attenuate activation of integrin-mediated arrest (Hafezi-Moghadam et al. 2001). Chemokine-accelerated, L-selectin-mediated rolling could thus serve to counterbalance excessive chemokine activation of integrin-mediated stoppage.

7.8 Endothelial Chemokines Attenuate L-Selectin Rolling by an Interference with Cytoskeletal Stabilization of Tethers

One of the most unexpected results of this study is the dependence of the chemokine downregulatory activity on the site density of the L-selectin ligand (Fig. 4 A; Grabovsky et al. 2002). Chemokine suppression of L-selectin-mediated rolling required a threshold density not only of chemokine but also of L-selectin ligand. The magnitude of L-selectin adhesion suppression was progressively reduced upon L-selectin ligand dilution (Fig. 4 A). Notably, at high dilution, the suppressive chemokine signal interfered with neither tether formation nor tether duration (Fig. 4 A). This suggested that GPCR occupancy by chemokine interferes with a multivalent L-selectin association with ligand. To further extend this possibility, we used an anti-L-selectin lectin domain (monoclonal antibody) mAb, DREG-200, as a surrogate for L-selectin ligand in order to verify that this dose dependence phenomenon is not restricted to L-selectin:carbohydrate bonds. Since this anti-L-selectin mAb binds L-selectin with orders of magnitude higher affinity than native carbohydrate L-selectin ligands (Dwir et al. 2000), L-selectin interactions with the mAb result in either immediate arrests or in transient tethers without subsequent rolling (Dwir et al. 2000). Despite these entirely different molecular properties of the mAb compared to ligand, different chemokines could still suppress L-selectin-mediated cell capture and arrest on high-density mAb (Fig. 4B). Consistent with suppression by chemokine of L-selectin multivalent binding events at local contact sites, the same suppressive chemokines failed to inhibit L-selectin tethering to the diluted anti-L-selectin mAb (Fig. 4B). Thus, regardless of whether L-selectin interacts with its native carbohydrate ligands or with mAb, occupancy of GPCRs by chemokines juxtaposed to L-selectin ligand or mAb, delivers a potent signal that interferes with a multivalent occupancy of L-selectin cluster at the subsecond adhesive contact.

Microkinectic analysis of individual lymphocyte tethers formed at a fixed shear stress on surfaces coated with a fixed density of the prototypic L-selectin ligand, GlyCAM-1 alone, or with the prototypic chemokine SLC was next performed to gain further insights into the kinetics of these suppressive events. The duration of L-selectin tethers which comprise leukocyte rolling motions is progressively shorter as these tethers are comprised of fewer adhesive bonds (Chen and Springer 1999). Thus, tether duration can serve as a sensitive reporter of effective L-selectin avidity at the contact zone (Fig. 1). Upon initial tethering to L-selectin ligand coated-surface, lymphocytes continued to roll on the ligand through engagement of successive reversible tethers. Over 80% of these tethers dissociated from the ligand with a first order dissociation rate corresponding to a $t_{1/2}$ of 19.5 ms. The remaining tethers lasted significantly longer, with a $t_{1/2}$ of 43 ms. Lymphocytes interacting with GlyCAM-1 in the presence of immobilized SLC failed to roll and engaged with the L-selectin ligand exclusively through short-lived tethers ($t_{1/2}$ of 17.8 ms). Thus, persistent PBL rolling on GlyCAM-1 is mediated by a subset of prolonged tethers with a $t_{1/2}$ >40 ms which make up about one fifth of all L-selectin-mediated tethers (Grabovsky et al. 2000).

The molecular basis of this subsecond destabilization of L-selectin is still unclear. Several nonmutually exclusive mechanisms could account for such destabilization. First, it may involve interference of occupied GPCRs with local extrusion of membrane projections (thin tethers) which is predicted to increase contact area and prolong its duration. However, extrusion of membrane tethers have been shown so far only for cells rolling on high-density P-selectin (Park et al. 2002; Schmidtke and Diamond 2000) and does not seem to occur in cells engaged through labile L-selectin interactions (O. Dwir and R. Alon, unpublished data). Another possibility is that chemokine-occupied GPCRs interfere with local stretching of microvilli which is predicted to reduce the rupture force on individual adhesive bonds and thereby prolong tether lifetime (Park et al. 2002; Shao et al. 1998). However, since tether lifetime to diluted L-selectin ligand is in fact not shortened by the chemokine signals (Fig. 4 A), where contribution of postulated stretching to reduced rupture force per bond and increased tether lifetime would be highest, this possibility is highly unlikely. Cell fixation which eliminates microvilli stretching also does not shorten P-selectin tether lifetime (Yago et al. 2002) or L-selectin tether lifetime (O. Dwir and R. Alon, unpublished data). Alternatively, selectin-mediated rolling may involve local deformation of microvilli (Yago et al. 2002). Such deformation could result in microvilli bending, which would increase the microvillar contact area and therefore the probability of bond formation (Chen and Springer 1999). Such stabilization would occur primarily at high density of ligand (or mAb to L-selectin) since at low density, ligand may be too sparse for multiple bond formation by bent microvilli. Interference of chemokine-occupied GPCRs with such postulated bending of individual microvilli could therefore result in inhibition of tether stabilized on high-density ligand (or mAb) but not on low-density ligand, as observed by us (Fig. 4 A, B). Determination of microvilli rigidity modulation upon subsecond encounter with immobilized chemokines will be necessary to test this attractive possibility. Another approach could be to test whether cytochalasins, known to increase microvilli deformability, can rescue chemokine suppression of L-selectin-mediated rolling. GPCR-induced suppression of microvillus deformation may involve chemokine-induced clustering of GPCRs juxtaposed to L-selectin (Grabovsky et al.

2002). Soluble chemokines fail to transduce destabilization signals to L-selectin, since they cannot cluster their GPCRs. Recent electron microscopic analysis of the chemokine receptors for SDF-1α and RANTES, CXCR4 and CCR5, respectively, in PBL has demonstrated that these GPCRs indeed localize to lymphocyte microvilli (Singer et al. 2001) possibly in close proximity to L-selectin (Bruehl et al. 1997; Picker et al. 1991; von Andrian et al. 1995). Whether other GPCRs like CCR7, CXCR5, and CXCR1/2, predicted to deliver potent L-selectin suppressive signals, are also found on leukocyte microvilli remains to be shown.

There is however, yet, another mechanism by which chemokine occupancy of GPCRS may inhibit L-selectin tether stabilization and suppress rolling. Suppressive chemokines may not only cluster their cognate GPCR molecules on the microvilli, but also promote their anchorage to the cytoskeleton. Rapid anchorage of GPCRs to the cytoskeleton nearby L-selectin may perturb the L-selectin association with the actin cytoskeleton necessary for immediate stabilization of nascent tethers (Dwir et al. 2001). In this respect, tail-truncated L-selectin with partially lost association with the actin cytoskeleton due to impaired binding to the actin linker, a-actinin, was still susceptible to suppression by chemokines (Grabovsky et al. 2002). This mechanism may therefore involve α-actinin-independent associations of L-selectin with cortical cytoskeletal components such as ERM proteins (Ivetic et al. 2002). It would be interesting to test whether L-selectin mutants abrogated in one or more of these associations are rendered resistant to chemokine-suppression of rolling.

7.9 Conclusions and Perspectives

This work raises the possibility that rolling adhesions which allow a captured leukocyte to sample the endothelium for specific chemokines are in fact subjected to a negative feedback mechanism by these very chemokines. Chemokine encounters by rolling leukocytes may serve to counterbalance the in situ activation of integrins by these encounters along the rolling path. This mechanism introduces a far larger flexibility and combinatorial complexity to the multistep process of leukocyte trafficking to target endothelia. Thus, the potential of a tethered leukocyte to respond to a given combination of traffic signals is not merely determined by proper expression of L-selectin ligands, integrin ligands, and chemokines on the endothelial target. Rather, the spatial distribution of these traffic signals on the endothelial surface may determine the efficiency by which leukocyte capture and rolling can lead to productive arrest on the target endothelial site. Chemokine destabilization of L-selectin adhesion may underlie the jerky nature of rolling mediated by L-selectin in vivo (Warnock et al. 1998). The jerky nature of L-selectin rolling is not controlled solely by chemokines. Antiadhesive glycoproteins like CD43 (Stockton et al. 1998), topological heterogeneity of both leukocyte and endothelial surfaces (Finger et al. 1996a), and the labile nature of individual L-selectin tethers depicted in vitro (Alon et al. 1997; Puri et al. 1997) can each contribute to this jerky nature observed in vivo. L-selectin rolling is also tightly regulated by the presence of shear forces (Chen and Springer 1999; Finger et al. 1996b; Lawrence et al. 1997). The existence of multiple mechanisms for modulating strength of L-selectin rolling further suggests that its dynamics and strength control the efficiency of subsequent integrin activation steps necessary for leukocyte arrest on the target endothelium. L-selectin-mediated rolling, in particular through sites expressing high levels of L-selectin ligands like peripheral lymph nodes (Maly et al. 1996; Rosen and Bertozzi 1994), should be tightly controlled to avoid L-selectin ligation events on rolling leukocytes which may directly activate integrins (Hwang et al. 1996; Simon et al. 1999; Steeber et al. 1997).

At least four distinct physiological scenarios may underlie this novel inhibitory mechanism of L-selectin rolling. For simplicity, the lymph node HEV and resting circulating lymphocytes could be taken as an example for such scenarios. Lymphocyte rolling on HEV regions expressing high-density L-selectin ligand and lymphocyte-specific chemokines is expected to be strongly suppressed. As a result, integrin ligand, if sufficiently high, can lead to rapid integrin activation by the high level chemokine resulting in immediate arrest, bypassing the loss of the rolling step. Indeed, subsets of lymphocytes can arrest on ICAM-1 post tethering through L-selectin interactions without intermediate rolling steps (Campbell et al. 1998). A second scenario may occur in HEV regions presenting high level L-selectin ligand but low levels of chemokines to a specific lymphocyte subset. L-selectin-mediated rolling rather than being abolished could become jerky and faster, resulting in reduced probability of integrin activation and arrest. A third and most interesting scenario is that chemokine (or chemokines) presented on HEV could bind GPCRs on L-selectin-expressing nonlymphoid cells without triggering Gi-stimulatory signals, through their GPCRs. Chemokines like SLC, ELC, or BCA-1 may exert suppressive signals through binding GPCRs on nonlymphoid leukocytes without activating their integrins. There are multiple examples of chemokines which bind with high affinity to a given GPCR but do not signal through it (Loetscher et al. 2001). These interactions may serve, for instance, to clear out from the lymph node vessels neutrophils and monocytes which roll on lymph node HEV but rarely arrest at these sites. In a related mechanism, native chemokines may clear out leukocytes expressing subsets of GPCRs which had been uncoupled from their Gi-protein machinery in cells subjected to anti-inflammatory cytokines (D'Amico et al. 2000). These GPCR subsets may not only restrict integrin activations but also suppress L-selectin-mediated rolling of these cells and thereby inhibit their recruitment to inflammatory sites expressing L-selectin ligands.

A fourth interesting scenario is the existence of chemokines on lymph node HEVs in partially truncated states with potent GPCR binding activity but poor signaling. For instance, SDF-1a, which is expressed at various zones of peripheral lymph nodes (Okada et al. 2002) can be truncated at its N' terminus by neutrophil secreted elastase or by serum MMPs (Valenzuela-Fernandez et al. 2001) generating a CXCR4 antagonist with no signaling capacity but significant binding activity (Crump et al. 1997). A related mutant has been shown by us to strongly suppress L-selectin rolling while lacking any integrin stimulatory activity (Grabovsky et al. 2000; Grabovsky et al. 2002). This and other partially truncated chemokines may bind their cognate GPCRs and thus suppress L-selectin-mediated rolling. This activity may counterbalance integrin stimulatory activities of the native nontruncated chemokine counterparts. Thus, inflammatory conditions may generate distinct chemokine derivatives acting as dominant suppressors of leukocyte recruitment at sites of L-selectin ligand activity.

Several key questions raised by these findings await future investigation. First, it is still unclear whether the mechanism of chemokine suppression identified by us is L-selectin specific. The possibility that it reflects interference with microvillar properties or with the cytoskeletal environment of microvilliar adhesion molecules makes it likely that rolling mediated by PSGL-1, a major ligand for P- and E-selectin (Moore et al. 1995; Xia et al. 2002) is also subject to chemokine suppression. In addition, specialization among GPCRs capable of activating integrins has been suggested in several cellular systems (Weber et al. 2001). Reminiscent of these findings, particular GPCRs or GPCR subsets, for instance receptors expressed in proximity to L-selectin, may be specialized to deliver suppressive signals to L-selectin-mediated rolling while other GPCRs excluded from microvilli should lack this activity. Future elucidation of this novel activity of chemokines should be valuable for the development of specific chemokine receptor antagonists with dual inhibitory activities, capable of attenuating both selectin-mediated rolling and activation of integrins in pathological processes of leukocyte recruitment.

Acknowledgements. We thank Drs. S. W. Feigelson for helpful discussions and S. Schwarzbaum for editorial assistance. R. Alon is the Incumbent of The Tauro Career Development Chair in Biomedical Research. The research discussed here has been supported by the Israel Science Foundation, the Minerva Foundation, Germany and the Pasteur-Weizmann Joint Research Program.

References

Alon R, Feizi T, Yuen CT, Fuhlbrigge RC, Springer TA (1995a) Glycolipid ligands for selectins support leukocyte tethering and rolling under physiologic flow conditions. J Immunol 154:5356–5366

Alon R, Hammer DA, Springer TA (1995b) Lifetime of the P-selectin-carbohydrate bond and its response to tensile force in hydrodynamic flow. Nature 374:539–542

Alon R, Chen S, Puri KD, Finger EB, Springer TA (1997) The kinetics of L-selectin tethers and the mechanics of selectin-mediated rolling. J Cell Biol 138:1169–1180

Alon R, Chen S, Fuhlbrigge R, Puri KD, Springer TA (1998) The kinetics and shear threshold of transient and rolling interactions of L-selectin with its ligand on leukocytes. Proc Natl Acad Sci USA 95:11631–11636

Bell G (1978) Models for the specific adhesion of cells to cells. Science 200:618–627

Bruehl RE, Moore KL, Lorant DE, Borregaard N, Zimmerman GA, McEver RP, Bainton DF (1997) Leukocyte activation induces surface redistribution of P-selectin glycoprotein ligand-1. J Leukoc Biol 61:489–499

Brzostowski JA, Kimmel AR (2001) Signaling at zero G: G-protein-independent functions for 7-TM receptors. Trends Biochem Sci 26:291–297

Butcher EC, Picker LJ (1996) Lymphocyte homing and homeostasis. Science 272:60–66

Campbell JJ, Hedrick J, Zlotnik A, Siani MA, Thompson DA (1998) Chemokines and the arrest of lymphocytes rolling under flow conditions. Science 279:381–384

Campbell JJ, Butcher EC (2000) Chemokines in tissue-specific and microenvironment-specific lymphocyte homing. Curr Opin Immunol 12:336–341

Chen C, Mobley JL, Dwir O, Shimron F, Grabovsky V, Lobb RL, Shimizu Y, Alon R (1999) High affinity VLA-4 subsets expressed on T cells are mandatory for spontaneous adhesion strengthening but not for rolling on VCAM-1 in shear flow. J Immunol 162:1084–1095

Chen S, Springer TA (1999) An automatic braking system that stabilizes leukocyte rolling by an increase in selectin bond number with shear. J Cell Biol 144:185–200

Chen S, Springer TA (2001) Selectin receptor-ligand bonds: Formation limited by shear rate and dissociation governed by the Bell model. Proc Natl Acad Sci USA 98:950–955

Crump MP, Gong JH, Loetscher P, Rajarathnam K, Amara A, Arenzana-Seisdedos F, Virelizier JL, Baggiolini M, Sykes BD, Clark-Lewis I (1997) Solution structure and basis for functional activity of stromal cell-derived factor-1; dissociation of CXCR4 activation from binding and inhibition of HIV-1. EMBO J 16:6996–7007

D'Amico G, Frascaroli G, Bianchi G, Transidico P, Doni A, Vecchi A, Sozzani S, Allavena P, Mantovani A (2000) Uncoupling of inflammatory chemokine receptors by IL-10: generation of functional decoys. Nat Immunol 1:387–391

Dwir O, Kansas GS, Alon R (2000) An activated L-selectin mutant with conserved equilibrium binding properties but enhanced ligand recognition under shear flow. J Biol Chem 275:18682–18691

Dwir O, Kansas GS, Alon R (2001) The cytoplasmic tail of L-selectin regulates leukocyte capture and rolling by controlling the mechanical stability of selectin:ligand tethers. J Cell Biol 155:145–156

Dwir O, Steeber DA, Schwarz US, Camphausen RT, Kansas GS, Tedder TF, Alon R (2002) L-selectin dimerization enhances tether formation to properly spaced ligand. J Biol Chem 277:21130–21139

Evans SS, Schleider DM, Bowman LA, Francis ML, Kansas GS, Black JD (1999) Dynamic association of L-selectin with the lymphocyte cytoskeletal matrix. J Immunol 162:3615–3624

Finger EB, Bruehl RE, Bainton DF, Springer TA (1996a) A differential role for cell shape in neutrophil tethering and rolling on endothelial selectins under flow. J Immunol 157:5085–5096

Finger EB, Puri KD, Alon R, Lawrence MB, von Andrian UH, Springer TA (1996b) Adhesion through L-selectin requires a threshold hydrodynamic shear. Nature 379:266–269

Grabovsky V, Feigelson S, Chen C, Bleijs R, Peled A, Cinamon G, Baleux F, Arenzana-Seisdedos F, Lapidot T, van Kooyk Y, Lobb R, Alon R (2000) Subsecond induction of $\alpha4$ integrin clustering by immobilized chemokines enhances leukocyte capture and rolling under flow prior to firm adhesion to endothelium. J Exp Med 192:495–505

Grabovsky V, Dwir O, Alon R (2002) Endothelial chemokines destabilize L-selectin-mediated lymphocyte rolling without inducing selectin shedding. J. Biol. Chem. 277:20640–20650

Greenberg AW, Brunk DK, Hammer DA (2000) Cell-free rolling mediated by L-selectin and sialyl lewis(x) reveals the shear threshold effect. Biophys J 79:2391–2402

Hafezi-Moghadam A, Ley K (1999) Relevance of L-selectin shedding for leukocyte rolling in vivo. J Exp Med 189:939–948

Hafezi-Moghadam A, Thomas KL, Prorock AJ, Huo Y, Ley K (2001) L-selectin shedding regulates leukocyte recruitment. J Exp Med 193:863–872

Hammer DA, Lauffenburger DA (1989) A dynamical model for receptor-mediated cell adhesion to surfaces in viscous shear flow. Cell Biophys 14:139–173

Hwang ST, Singer MS, Giblin PA, Yednock TA, Bacon KB, Simon SI, Rosen SD (1996) GlyCAM-1, a physiologic ligand for L-selectin, activates $\beta2$ integrins on naive peripheral lymphocytes. J Exp Med 184:1343–1348

Ivetic A, Deka J, Ridley A, Ager A (2002) The cytoplasmic tail of L-selectin interacts with members of the Ezrin-Radixin-Moesin (ERM) family of proteins: cell activation-dependent binding of Moesin but not Ezrin. J Biol Chem 277:2321–2329

Kansas GS, Ley K, Munro JM, Tedder TF (1993) Regulation of leukocyte rolling and adhesion to high endothelial venules through the cytoplasmic domain of L-selectin. J Exp Med 177:833–838

Kansas GS, Saunders KB, Ley K, Zakrzewicz A, Gibson RM, Furie BC, Furie B, Tedder TF (1994) A role for the epidermal growth factor-like domain of P-selectin in ligand recognition and cell adhesion. J Cell Biol 124:609–618

Kansas GS (1996) Selectins and their ligands: current concepts and controversies. Blood. 88:3259–3287

Kaplanski G, Farnarier C, Tissot O, Pierres A, Benoliel AM, Alessi MC, Kaplanski S, Bongrand P (1993) Granulocyte-endothelium initial adhesion. Analysis of transient binding events mediated by E-selectin in a laminar shear flow. Biophys J 64:1922–1933

Lasky LA (1991) Lectin cell adhesion molecules (LEC-CAMs): a new family of cell adhesion proteins involved with inflammation. J Cell Biochem 45:139–146

Lawrence MB, Springer TA (1991) Leukocytes roll on a selectin at physiologic flow rates: distinction from and prerequisite for adhesion through integrins. Cell 65:859–873

Lawrence MB, Springer TA (1993) Neutrophils roll on E-selectin. J Immunol 151:6338–6346

Lawrence MB, Kansas GS, Kunkel EJ, Ley K (1997) Threshold levels of fluid shear promote leukocyte adhesion through selectins (CD62L,P,E). J Cell Biol 136:717–727

Li X, Steeber DA, Tang MLK, Farrar MA, Perlmutter RM, Tedder TF (1998) Regulation of L-selectin-mediated rolling through receptor dimerization. J Exp Med 188:1385–1390

Loetscher P, Pellegrino A, Gong JH, Mattioli I, Loetscher M, Bardi G, Baggiolini M, Clark-Lewis I (2001) The ligands of CXC chemokine receptor 3, I-TAC, Mig, and IP10, are natural antagonists for CCR3. J Biol Chem 276:2986–2991

Mackay CR (2001) Chemokines: immunology's high impact factors. Nat Immunol 2:95–101

Maly P, Thall A, Petryniak B, Rogers CE, Smith PL, Marks RM, Kelly RJ, Gersten KM, Cheng G, Saunders TL, Camper SA, Camphausen RT, Sullivan FX, Isogai Y, Hindsgaul O, von Andrian UH, Lowe JB (1996) The Fuc-TVII α1,3 fucosyltransferase controls leukocyte trafficking through an essential role in L-, E-, and P-selectin ligand biosynthesis. Cell 86:643–653

Mazo IB, von Andrian UH (1999) Adhesion and homing of blood-borne cells in bone marrow microvessels. J Leuko Biol 66:25–32

Mehta P, Cummings RD, McEver RP (1998) Affinity and kinetic analysis of P-selectin binding to P-selectin glycoprotein ligand-1. J Biol Chem 273:32506–32513

Moore KL, Eaton SF, Lyons DE, Lichenstein HS, Cummings RD, McEver RP (1994) The P-selectin glycoprotein ligand from human neutrophils displays sialylated, fucosylated, O-linked poly-N-acetyllactosamine. J Biol Chem 269:23318–23327

Moore KL, Patel KD, Bruehl RE, Fungang L, Johnson DL, Lichenstein HS, Cummings RD, Bainton DF, McEver RP (1995) P-selectin glycoprotein ligand-1 mediates rolling of human neutrophils on P- selectin. J Cell Biol 128:661–671

Muller A, Homey B, Soto H, Ge N, Catron D, Buchanan ME, McClanahan T, Murphy E, Yuan W, Wagner SN, Barrera JL, Mohar A, Verastegui E, Zlotnik A (2001) Involvement of chemokine receptors in breast cancer metastasis. Nature 410:50–56

Nicholson MW, Barclay AN, Singer MS, Rosen SD, van der Merwe PA (1998) Affinity and kinetic analysis of L-selectin binding to GlyCAM-1. J Biol Chem 273:763–770

Okada T, Ngo VN, Ekland EH, Forster R, Lipp M, Littman DR, Cyster JG (2002) Chemokine requirements for B cell entry to lymph nodes and Peyer's patches. J Exp Med 196:65–75

Park EY, Smith MJ, Stropp ES, Snapp KR, DiVietro JA, Walker WF, Schmidtke DW, Diamond SL, Lawrence MB (2002) Comparison of PSGL-1 microbead and neutrophil rolling: microvillus elongation stabilizes p-selectin bond clusters. Biophys J 82:1835–1847

Patel KD, Nollert MU, McEver RP (1995) P-selectin must extend a sufficient length from the plasma membrane to mediate rolling of neutrophils. J Cell Biol 131:1893–1902

Pavalko FM, Walker DM, Graham L, Goheen M, Doerschuk CM, Kansas GS (1995) The cytoplasmic domain of L-selectin interacts with cytoskeletal proteins via α-actinin:receptor positioning in microvilli does not require interaction with α-actinin. J Cell Biol 129:1155–1164

Peled A, Petit I, Kollet O, Magid M, Ponomaryov T, Byk T, Nagler A, Ben-Hur H, Many A, Shultz L, Lider O, Alon R, Zipori D, Lapidot T (1999) Dependence of human stem cell engraftment and repopulation of NOD/SCID mice on CXCR4. Science 283:845–848

Picker LJ, Warnock RA, Burns AR, Doerschuk CM, Berg EL, Butcher EC (1991) The neutrophil selectin LECAM-1 presents carbohydrate ligands to the vascular selectins ELAM-1 and GMP-140. Cell 66:921–933

Puri KD, Springer TA (1996) A schiff base with mildly oxidized carbohydrate ligands stabilizes L-selectin and not P-selectin or E-selectin rolling adhesions in shear flow. J Biol Chem 271:5404–5413

Puri KD, Finger EB, Springer TA (1997) The faster kinetics of L-selectin than of E-selectin and P-selectin rolling at comparable binding strength. J Immunol 158:405–413

Puri KD, Chen S, Springer TA (1998) Modifying the mechanical property and shear threshold of L-selectin adhesion independently of equilibrium properties. Nature. 392:930–933

Ramachandran V, Nollert MU, Qiu H, Liu WJ, Cummings RD, Zhu C, McEver RP (1999) Tyrosine replacement in P-selectin glycoprotein ligand-1 affects distinct kinetic and mechanical properties of bonds with P- and L-selectin. Proc Natl Acad Sci USA 96:13771–13776

Ramachandran V, Yago T, Epperson TK, Kobzdej MM, Nollert MU, Cummings RD, Zhu C, McEver RP (2001) Dimerization of a selectin and its ligand stabilizes cell rolling and enhances tether strength in shear flow. Proc Natl Acad Sci USA 98:10166–10171

Rosen SD, Bertozzi CR (1994) The selectins and their ligands. Curr Opin Cell Biol 6:663–673

Schmidtke DW, Diamond SL (2000) Direct observation of membrane tethers formed during neutrophil attachment to platelets or P-selectin under physiological flow. J Cell Biol 149:719–729

Setiadi H, Sedgewick G, Erlandsen SL, McEver RP (1998) Interactions of the cytoplasmic domain of P-selectin with clathrin-coated pits enhance leukocyte adhesion under flow. J Cell Biol 142:859–871

Shamri R, Grabovsky V, Feigelson S, Dwir O, Van Kooyk Y, Alon R (2002) Chemokine-stimulation of lymphocyte $\alpha4$ integrin avidity but not of LFA-1 avidity to endothelial ligands under shear flow requires cholesterol membrane rafts. J Biol Chem (in press)

Shao JY, Ting-Beall HP, Hochmuth RM (1998) Static and dynamic lengths of neutrophil microvilli. Proc Natl Acad Sci USA 95:6797–6802

Shao JY, Hochmuth RM (1999) Mechanical anchoring strength of L-selectin, $\beta2$ integrins, and CD45 to neutrophil cytoskeleton and membrane. Biophys J 77:587–596

Shimizu Y, Rose DM, Ginsberg MH (1999) Integrins in the immune system. Adv Immunol 72:325–380

Simon SI, Cherapanov V, Nadra I, Waddell TK, Seo SM, Wang Q, Doerschuk CM, Downey GP (1999) Signaling functions of L-selectin in neutrophils: alterations in the cytoskeleton and colocalization with CD18. J Immunol 163:2891–2901

Singer II, Scott S, Kawka DW, Chin J, Daugherty BL, DeMartino JA, DiSalvo J, Gould SL, Lineberger JE, Malkowitz L, Miller MD, Mitnaul L, Siciliano SJ, Staruch MJ, Williams HR, Zweerink HJ, Springer MS (2001) CCR5, CXCR4, and CD4 are clustered and closely apposed on microvilli of human macrophages and T cells. J Virol 75:3779–3790

Smith MJ, Berg EL, Lawrence MB (1999) A direct comparison of selectin-mediated transient, adhesive events using high temporal resolution. Biophys J 77:3371–3383

Snapp KR, Craig R, Herron M, Nelson RD, Stoolman LM, Kansas GS (1998) Dimerization of P-selectin glycoprotein ligand-1 (PSGL-1) required for optimal recognition of P-selectin. J Cell Biol 142:263–270

Snapp KR, Heitzig CE, Kansas GS (2002) Attachment of the PSGL-1 cytoplasmic domain to the actin cytoskeleton is essential for leukocyte rolling on P-selectin. Blood 99:4494–4502

Somers WS, Tang J, Shaw GD, Camphausen RT (2000) Insights into the molecular basis of leukocyte tethering and rolling revealed by structures of P- and E-selectin bound to sLe(x) and PSGL-1. Cell 103:467–479

Springer TA (1994) Traffic signals for lymphocyte recirculation and leukocyte emigration: The multistep paradigm. Cell 76:301–314

Steeber DA, Engel P, Miller AS, Sheetz MP, Tedder TF (1997) Ligation of L-selectin through conserved regions within the lectin domain activates signal transduction pathways and integrin function in human, mouse, and rat leukocytes. J Immunol 159:952–963

134   O. Dwir et al.

Stein JV, Rot A, Luo Y, Narasimhaswamy M, Nakano H, Gunn MD, Matsu-
zawa A, Quackenbush EJ, Dorf ME, von Andrian UH (2000) The CC
chemokine thymus-derived chemotactic agent 4 (TCA-4, secondary lymphoid tissue chemokine, 6Ckine, Exodus-2) triggers lymphocyte function-
associated antigen 1-mediated arrest of rolling T lymphocytes in peripheral lymph node high endothelial venules. J Exp Med 191:61–76
Stockton BM, Cheng G, Manjunath N, Ardman B, von Andrian UH (1998)
Negative regulation of T cell homing by CD43. Immunity 8:373–381
Thelen M (2001) Dancing to the tune of chemokines. Nat Immunol 2:129–
134
Valenzuela-Fernandez A, Palanche T, Amara A, Magerus A, Altmeyer R,
Delaunay T, Virelizier JL, Baleux F, Galzi JL, Arenzana-Seisdedos FF
(2001) Optimal inhibition of X4 HIV isolates by the CXC chemokine
stromal cell-derived factor 1 alpha requires interaction with cell surface
heparan sulfate proteoglycans. J Biol Chem 276:26550–26558
van der Merwe PA (1999) Leukocyte adhesion: High-speed cells with ABS.
Curr Biol 9:R419-R422
Vila-Coro AJ, Rodriguez-Frade JM, Martin De Ana A, Moreno-Ortiz MC,
Martinez AC, Mellado M (1999) The chemokine SDF-1α triggers
CXCR4 receptor dimerization and activates the JAK/STAT pathway.
FASEB J 13:1699–1710
von Andrian UH, Hasslen SR, Nelson RD, Erlandsen SL, Butcher EC
(1995) A central role for microvillous receptor presentation in leukocyte
adhesion under flow. Cell 82:989–999
Walcheck B, Kahn J, Fisher JM, Wang BB, Fisk RS, Payan DG, Feehan C,
Betageri R, Darlak K, Spatola AF, Kishimoto TK (1996) Neutrophil roll-
ing altered by inhibition of L-selectin shedding in vitro. Nature 380:720–
723
Warnock RA, Askari S, Butcher EC, von Andrian UH (1998) Molecular
mechanisms of lymphocyte homing to peripheral lymph nodes. J Exp
Med 187:205–216
Weber C, Weber KS, Klier C, Gu S, Wank R, Horuk R, Nelson PJ (2001)
Specialized roles of the chemokine receptors CCR1 and CCR5 in the re-
cruitment of monocytes and Th1-like/CD45RO$^+$ T cells. Blood 97:1144–
1146
Wild MK, Huang MC, Schulze-Horsel U, van der Merwe PA, Vestweber D
(2001) Affinity, kinetics, and thermodynamics of E-selectin binding to E-
selectin ligand-1. J Biol Chem 276:31602–31612
Xia L, Sperandio M, Yago T, McDaniel JM, Cummings RD, Pearson-White
S, Ley K, McEver RP (2002) P-selectin glycoprotein ligand-1-deficient
mice have impaired leukocyte tethering to E-selectin under flow. J Clin
Invest 109:939–950
Yago T, Leppanen A, Qiu H, Marcus WD, Nollert MU, Zhu C, Cummings
RD, McEver RP (2002) Distinct molecular and cellular contributions to
stabilizing selectin-mediated rolling under flow. J Cell Biol 158:787–799

Yeh JC, Hiraoka N, Petryniak B, Nakayama J, Ellies LG, Rabuka D, Hindsgaul O, Marth JD, Lowe JB, Fukuda M (2001) Novel sulfated lymphocyte homing receptors and their control by a Core1 extension beta 1,3-N-acetylglucosaminyltransferase. Cell 105:957–969

8 Interactions of Selectins with PSGL-1 and Other Ligands

R. P. McEver

The selectins are type I membrane glycoproteins that mediate adhesion of leukocytes and platelets to vascular surfaces (McEver 2001; Vestweber and Blanks 1999). L-selectin is expressed on most leukocytes. E-selectin is expressed on cytokine-activated endothelial cells. P-selectin is rapidly redistributed from membranes of secretory granules to the surfaces of activated platelets and endothelial cells. Each selectin has a membrane-distal C-type lectin domain, followed by an epidermal growth factor (EGF)-like motif, a series of consensus repeats, a transmembrane domain, and a short cytoplasmic tail. P- and E-selectin bind primarily to ligands on leukocytes, and P-selectin also interacts with ligands on platelets and some endothelial cells. L-selectin binds to ligands on endothelial cells of high endothelial venules (HEV) in lymph nodes and on other leukocytes.

8.1 Selectin Ligands

Like most lectins, selectins bind to a range of glycoconjugates with varying affinities. Key challenges have been to identify preferred glycoconjugates with higher affinity or avidity for selectins, to elucidate the biosynthetic pathways for selectin ligands, and to determine which glycoconjugates actually mediate cell adhesion to selectins under physiological flow. All selectins bind with low affinity to glycans with terminal components that include α2,3-linked sialic acid and α1,3-linked fucose, typified by the sialyl Lewis x (sLe^x) determinant (NeuAcα2,3Galβ1,3[Fucα1,3]GlcNAcβ1-R). Crystal structures of sLe^x bound to the lectin domains of P- and E-selectin reveal a network of interactions between the fucose, a single Ca^{2+} ion, and several amino acids, including those that coordinate the Ca^{2+}; this explains the Ca^{2+}-dependent nature of binding to fucosylated glycans (Somers et al. 2000). The sialic acid and the galactose also interact with the lectin domain. Targeted disruption of the gene encoding the α1,3 fucosyltransferase Fuc-TVII in mice significantly decreases selectin-mediated leukocyte trafficking, and disruption of the genes for both Fuc-TVII and Fuc-TIV eliminates these interactions (Homeister et al. 2001; Smithson et al. 2001). These studies suggest that virtually all physiologically relevant selectin ligands require α1,3-linked fucose.

P- and L-selectin, but not E-selectin, also bind in a Ca^{2+}-independent manner to sulfated glycans such as heparin, fucoidan, and sulfoglucuronyl glycolipids, which lack sialic acid or fucose (Varki 1997). These interactions suggested that sulfation of fucosylated glycoconjugates might enhance their interactions with P- and L-selectin. Subsequent studies confirmed this notion and documented that sulfation may occur either on tyrosines or on the glycans of a glycoprotein. L-selectin binds to a series of mucins expressed by HEV of lymph nodes. These mucins have many core-2 O-glycans capped with sLe^x. The mucins are sulfated on the C6 position of galactose and N-acetylglucosamine (GlcNAc) residues on numerous core-2 or extended core-1 O-glycans (Hemmerich et al. 2001; Yeh et al. 2001). Conceivably, the sLe^x and sulfate components could be contributed by a single core-2 branch or a single core-1 extension. Alternatively, the sLe^x and the sulfate could be separately contributed by a core-2 branch and core-1 extension on a biantennary O-glycan or by two clustered O-glycans. The precise structural basis for how these various components cooperate to optimize binding to L-selectin is not known.

The selectin ligand with the most clearly demonstrated biological functions is P-selectin glycoprotein ligand-1 (PSGL-1), which is expressed on leukocytes (McEver 2001; McEver and Cummings 1997). PSGL-1 is a transmembrane, homodimeric mucin bearing multiple *O*-glycans on serines and threonines. Antibody blocking studies and genetic deletion of PSGL-1 demonstrate that PSGL-1 is the dominant ligand for P-selectin and L-selectin on leukocytes. Studies with synthetic glycosulfopeptides indicate that P-selectin binds in a stereospecific manner to the N-terminal region of human PSGL-1 through recognition of tyrosine sulfate residues, adjacent peptide determinants, and fucose, galactose, and sialic residues on a properly positioned core-2 *O*-glycan (Leppänen et al. 1999; Leppänen et al. 2000). The glycosulfopeptide must present sLe^x on a short core-2 *O*-glycan. Human PSGL-1 has a small number of these short, fucosylated *O*-glycans, but has more *O*-glycans bearing sLe^x on an extended, polyfucosylated core-2 branch (Wilkins et al. 1996). However, a glycosulfopeptide with this extended glycan binds poorly to P-selectin (Leppänen et al. 2002). Thus, optimal binding of P-selectin is critically dependent on the relative orientations of sLe^x and the sulfated peptide. The crystal structure of P-selectin complexed with a PSGL-1-derived glycosulfopeptide with sLe^x on a short core-2 *O*-glycan reveals a broad, shallow binding interface (Somers et al. 2000). The Ca^{2+}-dependent interactions with sLe^x on the core-2 *O*-glycan are supplemented by Ca^{2+}-independent contacts with tyrosine sulfate and other amino acids. These additional contacts explain why P-selectin binds with much higher affinity to PSGL-1 than to sLe^x alone. Targeted deletion of the murine gene encoding Core2GlcNAcT-I, the major core-2 β1,6-*N*-acetylglucosaminyltransferase in leukocytes, eliminates binding of leukocytes to P-selectin (Ellies et al. 1998; Sperandio et al. 2001). This suggests that Core2GlcNAcT-I plays a key role in constructing the relevant core-2 *O*-glycan on PSGL-1.

Like human PSGL-1, murine PSGL-1 is a homodimeric sialomucin (Lenter et al. 1994; Yang et al. 1996). mAbs to N-terminal peptide epitopes of PSGL-1 block rolling of murine leukocytes on murine P-selectin in vitro and in vivo (Borges et al. 1997a; Borges et al. 1997b; Steegmaier et al. 1997). This suggests that P-selectin probably also binds to the N terminus of murine PSGL-1. However, the

N terminus of murine PSGL-1 has a very different amino acid sequence than human PSGL-1 (Yang et al. 1996). Murine PSGL-1 has potential sites for sulfation at Tyr-13 and -15 and for O-glycosylation at Thr-14 and -17. Site-directed mutagenesis studies suggest that murine PSGL-1 requires sulfation of Tyr-13 and O-glycosylation of Thr-17 to bind optimally to P-selectin (Xia et al. 2003). Because it uses only one tyrosine, murine PSGL-1 may rely more on other peptide components and O-glycosylation to bind to P-selectin than does human PSGL-1.

Enzymatic desialylation of murine leukocytes eliminates binding to E- and P-selectin (Varki 1994), and selectins do not bind to leukocytes from mice that are deficient in Fuc-TVII and Fuc-TIV, the α1,3-fucosyltransferases normally expressed in these cells (Maly et al. 1996; Weninger et al. 2000). These combined data suggest that selectins recognize sLe^x-related glycans on murine leukocytes. However, many mAbs to sLe^x and Le^x fail to bind to murine leukocytes (Ito et al. 1994; Thorpe and Feizi 1984). It has been widely assumed that unknown glycan modifications, perhaps unique to murine tissues, mask the epitopes for these mAbs. This assumption has not been critically tested, since direct structural characterization of glycans on murine leukocytes has not been performed. Furthermore, treatment of murine monocytic WEHI-3 cells with glycosidases or chlorate demonstrated that sialic acid modifications, α1,3-galactosylation, or sulfation do not mask epitopes for mAbs to sLe^x or Le^x (Kobzdej et al. 2002). WEHI-3 cells and murine neutrophils express low α1,3-fucosyltransferase activities compared to human promyelocytic HL-60 cells. Consistent with very low endogenous fucosylation, forced fucosylation of intact WEHI-3 cells or murine neutrophils by exogenous α1,3-fucosyltransferase Fuc-TVI and GDP-fucose creates many new epitopes for anti-sLe^x mAbs such as HECA-452 and CSLEX-1. Nevertheless, forced fucosylation of intact cells does not significantly augment their ability to bind to fluid-phase P- or E-selectin or to roll on immobilized P- or E-selectin under flow (Kobzdej et al. 2002). These data suggest that murine myeloid leukocytes fucosylate only a few specific glycans, which interact preferentially with P- and E-selectin. Even on human HL-60 cells, which express abundant sLe^x determinants, only a small minority of O-glycans on PSGL-1 are fucosylated, and perhaps only one or two of these O-glycans has the short core-2 O-glycan capped with sLe^x that cooperates with sulfated tyrosines and peptide components to bind optimally to P-selectin (Leppänen et al. 2002; Wilkins et al. 1996). Therefore, both human and murine leukocytes may employ limited but specific α1,3-fucosylation to synthesize glycoconjugates that bind to selectins.

Identifying physiologically relevant glycoprotein ligands for E-selectin has been particularly difficult because most cells roll on E-selectin if they are transfected with an expression vector encoding an α1,3-fucosyltransferase that introduces sLe^x epitopes on the cell surface. Thus, a frequent consideration is that E-selectin interacts indiscriminately with sLe^x-terminated glycans on many glycoproteins or glycolipids on leukocytes. However, genetic deletion of PSGL-1 in mice impairs leukocyte tethering to E-selectin in vitro and in vivo (Xia et al. 2002). The residual leukocytes that tether roll equivalently to wild-type leukocytes. This demonstrates a novel function of PSGL-1 in tethering free-flowing leukocytes to E-selectin, but not in stabilizing subsequent rolling. A similar phenotype is seen in Core2GlcNAcT-I-deficient mice, suggesting that the sLe^x-capped core-2 O-glycans that interact with E-selectin are primarily on PSGL-1 (Sperandio et al. 2001). These data establish PSGL-1 as a physiologically relevant glycoprotein ligand for all three selectins.

8.2 Regulation of Cell Rolling Under Flow

Rolling cell adhesion in the vasculature requires the rapid formation and breakage of adhesive bonds that are subjected to applied force (McEver 2001). Surface plasmon resonance measurements indicate that the association and dissociation kinetics of unstressed selectin-ligand bonds are rapid, although they vary considerably depending on the selectin and the ligand (Mehta et al. 1998; Nicholson et al. 1998; Wild et al. 2001). The lifetimes of transient leukocyte tethers on low-density selectins shorten in response to increasing wall shear stress. This is a characteristic of "slip bonds," where tensile force accelerates dissociation. However, very low tensile forces prolong lifetimes of P-selectin/PSGL-1 bonds, as measured by both atomic force microscopy and transient tether lifetimes (Marshall et al. 2003). This counterintuitive behavior is a characteristic of "catch bonds," which have not been demonstrated previously. Transitions between catch and slip bonds may help explain the requirement for a minimum wall shear stress below which cells do not roll on selectins. Bonds experience increasing forces as they move from the leading edge to the trailing edge of the rolling cell, potentially shifting from the catch bond regime to the slip bond regime. Catch bonds may act to prevent premature dissociation before the bonds arrive at the trailing edge, thereby maintaining stable rolling.

Leukocytes roll on selectins at nearly constant velocities over a wide range of wall shear stresses (Chen and Springer 1999). This "automatic braking system" has been ascribed to intrinsic molecular features of selectins and their ligands; higher wall shear stresses are postulated to overcome repulsive forces and increase bond formation. However, microspheres coupled with selectin ligands do not roll stably on selectins, whereas the same ligands coupled to cell surfaces confer stable rolling over a wide range of wall shear stresses (Yago et al. 2002). Fixation of the cells before ligand coupling destabilizes rolling, as observed for ligand-coupled microspheres. This indicates that stable selectin-dependent rolling requires cellular features as well as the intrinsic molecular components of selectins and their ligands. These features may include cellular deformation, which increases the adhesive contact area (Lei et al. 1999), and extrusion of long membrane tethers, which reduces the force on tethers and allows the cell to slip or roll downstream from the tether (Schmidtke and Diamond 2000).

The organization and orientation of selectins and their ligands on cell surfaces play major roles in modulating adhesion under flow (McEver 2001). Dimerization of P-selectin and PSGL-1 has little effect on the initial tethering of cells but does stabilize subsequent rolling (Ramachandran et al. 2001). If a cell is tethered by interactions of a dimeric selectin with a dimeric ligand, one bond may dissociate, leaving the cell still tethered. The first bond may then form again, stabilizing the tether and prolonging its lifetime. Replacement of the EGF domain of L-selectin with that of P-selectin alters the orientation of the lectin domain of L-selectin, allowing it to bind ligands more rapidly under flow. The result is more effective tethering and more stable rolling (Dwir et al. 2000).

Clustering of selectins or selectin ligands through interactions with cytoskeletal proteins provides an additional mechanism to increase bond number, thereby reducing the force on individual bonds and prolonging tether lifetimes (McEver 2001). The cytoplasmic domain of L-selectin contains a membrane-distal binding site for α-actinin (Pavalko et al. 1995) and a membrane-proximal binding site for moesin (Ivetic et al. 2002). Truncation of the region that includes the α-actinin binding site destabilizes rolling by shortening the lifetimes of adhesive tethers under flow (Dwir et al. 2001). Deletion of both the moesin and α-actinin binding sites virtually eliminates rolling, suggesting that both adaptors cooperate to link L-selectin to a stable network of actin filaments. L-selectin lacking the α-actinin binding site but retaining the moesin binding site is concentrated in microvilli like the wild-type protein (Kansas et al. 1993). Together, these data suggest that moesin bridges L-selectin and actin filaments in microvilli. The cytoplasmic domain of PSGL-1, which is also concentrated in microvilli, contains a membrane-proximal binding site for moesin (Serrador et al. 2002). Truncation of the cytoplasmic domain eliminates rolling of some PSGL-1-expressing cells on P-selectin (Snapp et al. 2002). However, the rolling defect is less severe in other assays (A.G. Klopocki, V. Ramachandran, R.P. McEver, unpublished data). It is possible that the cytoplasmic domain of PSGL-1 binds to more than one cytoskeletal adaptor, as does L-selectin. Further study is required to determine the contributions of specific portions of the cytoplasmic domains of L-selectin and PSGL-1 to their cell-surface organizations and adhesive functions.

References

Borges E, Eytner R, Moll T, Steegmaier M, Campbell MA, Ley K, Mossman H, Vestweber D (1997a) The P-selectin glycoprotein ligand-1 is important for recruitment of neutrophils into inflamed mouse peritoneum. Blood 90:1934–1942
Borges E, Tietz W, Steegmaier M, Moll T, Hallmann R, Hamann A, Vestweber D (1997b) P-selectin glycoprotein ligand-1 (PSGL-1) on T helper 1 but not on T helper 2 cells binds to P-selectin and supports migration into inflamed skin. J Exp Med 185:573–578

Chen SQ, Springer TA (1999) An automatic braking system that stabilizes leukocyte rolling by an increase in selectin bond number with shear. J Cell Biol 144:185–200

Dwir O, Kansas GS, Alon R (2000) An activated L-selectin mutant with conserved equilibrium binding properties but enhanced ligand recognition under shear flow. J Biol Chem 275:18682–18691

Dwir O, Kansas GS, Alon R (2001) Cytoplasmic anchorage of L-selectin controls leukocyte capture and rolling by increasing the mechanical stability of the selectin tether. J Cell Biol 155:145–156

Ellies LG, Tsuboi S, Petryniak B, Lowe JB, Fukuda M, Marth JD (1998) Core 2 oligosaccharide biosynthesis distinguishes between selectin ligands essential for leukocyte homing and inflammation. Immunity 9:881–890

Hemmerich S, Bistrup A, Singer MS, van Zante A, Lee JK, Tsay D, Peters M, Carminati JL, Brennan TJ, Carver-Moore K, Leviten M, Fuentes ME, Ruddle NH, Rosen SD (2001) Sulfation of L-selectin ligands by an HEV-restricted sulfotransferase regulates lymphocyte homing to lymph nodes. Immunity 15:237–247

Homeister JW, Thall AD, Petryniak B, Maly P, Rogers CE, Smith PL, Kelly RJ, Gersten KM, Askari SW, Cheng GY, et al. (2001) The α(1,3)fucosyltransferases FucT-IV and FucT-VII exert collaborative control over selectin-dependent leukocyte recruitment and lymphocyte homing. Immunity 15:115–126

Ito K, Handa K, Hakomori S (1994) Species-specific expression of sialosyl-Le(x) on polymorphonuclear leukocytes (PMN), in relation to selectin-dependent PMN responses. Glycoconj J 11:232–237

Ivetic A, Deka J, Ridley A, Ager A (2002) The cytoplasmic tail of L-selectin interacts with members of the ezrin-radixin-moesin (ERM) family of proteins. J Biol Chem 277:2321–2329

Kansas GS, Ley K, Munro JM, Tedder TF (1993) Regulation of leukocyte rolling and adhesion to high endothelial venules through the cytoplasmic domain of L-selectin. J Exp Med 177:833–838

Kobzdej MMA, Leppänen A, Ramachandran V, Cummings RD, McEver RP (2002) Discordant expression of selectin ligands and sialyl Lewis x-related epitopes on murine myeloid cells. Blood 100:485–494

Lei X, Lawrence MB, Dong C (1999) Influence of cell deformation on leukocyte rolling adhesion in shear flow. J Biomech Eng 121:636–643

Lenter M, Levinovitz A, Isenmann S, Vestweber D (1994) Monospecific and common glycoprotein ligands for E- and P-selectin on myeloid cells. J Cell Biol 125:471–481

Leppänen A, Mehta P, Ouyang Y-B, Ju T, Helin J, Moore KL, van Die I, Canfield WM, McEver RP, Cummings RD (1999) A novel glycosulfopeptide binds to P-selectin and inhibits leukocyte adhesion to P-selectin. J Biol Chem 274:24838–24848

Leppänen A, Penttilä L, Renkonen O, McEver RP, Cummings RD (2002) Glycosulfopeptides with O-glycans containing sialylated and polyfucosylated polylactosamine bind with low affinity to P-selectin. J Biol Chem 277:39749–39759

Leppänen A, White SP, Helin J, McEver RP, Cummings RD (2000) Binding of glycosulfopeptides to P-selectin requires stereospecific contributions of individual tyrosine sulfate and sugar residues. J Biol Chem 275:39569–39578

Maly P, Thall AD, Petryniak B, Rogers GE, Smith PL, Marks RM, Kelly RJ, Gersten KM, Cheng GY, Saunders TL, et al. (1996) The $\alpha(1,3)$Fucosyltransferase Fuc-TVII controls leukocyte trafficking through an essential role in L-, E-, and P-selectin ligand biosynthesis. Cell 86:643–653

Marshall BT, Long M, Piper JW, Yago T, McEver RP, Zhu C (2003) Direct observation of catch bonds involving cell-adhesion molecules. Nature 423:190–193

McEver RP (2001) Adhesive interactions of leukocytes, platelets, and the vessel wall during hemostasis and inflammation. Thromb Haemost 86:746–756

McEver RP, Cummings RD (1997) Role of PSGL-1 binding to selectins in leukocyte recruitment. J Clin Invest 100:485–492

Mehta P, Cummings RD, McEver RP (1998) Affinity and kinetic analysis of P-selectin binding to P-selectin glycoprotein ligand-1. J Biol Chem 273:32506–32513

Nicholson MW, Barclay AN, Singer MS, Rosen SD, Van der Merwe PA (1998) Affinity and kinetic analysis of L-selectin (CD62L) binding to glycosylation-dependent cell-adhesion molecule-1. J Biol Chem 273:763–770

Pavalko FM, Walker DM, Graham L, Goheen M, Doerschuk CM, Kansas GS (1995) The cytoplasmic domain of L-selectin interacts with cytoskeletal proteins via α-actinin: Receptor positioning in microvilli does not require interaction with α-actinin. J Cell Biol 129:1155–1164

Ramachandran V, Yago T, Epperson TK, Kobzdej MMA, Nollert MU, Cummings RD, Zhu C, McEver RP (2001) Dimerization of a selectin and its ligand stabilizes cell rolling and enhances tether strength in shear flow. Proc Natl Acad Sci USA 98:10166–10171

Schmidtke DW, Diamond SL (2000) Direct observation of membrane tethers formed during neutrophil attachment to platelets or P-selectin under physiological flow. J Cell Biol 149:719–729

Serrador JM, Urzainqui A, Alonso-Lebrero JL, Cabrero JR, Montoya MC, Vicente-Manzanares M, Yanez-Mo M, Sanchez-Madrid F (2002) A juxtamembrane amino acid sequence of P-selectin glycoprotein ligand-1 is involved in moesin binding and ezrin/radixin/moesin-directed targeting at the trailing edge of migrating lymphocytes. Eur J Immunol 32:1560–1566

Smithson G, Rogers CE, Smith PL, Scheidegger EP, Petryniak B, Myers JT, Kim DSL, Homeister JW, Lowe JB (2001) Fuc-TVII is required for T helper 1 and T cytotoxic 1 lymphocyte selectin ligand expression and recruitment in inflammation, and together with Fuc-TIV regulates naive T cell trafficking to lymph nodes. J Exp Med 194:601–614

Snapp KR, Heitzig CE, Kansas GS (2002) Attachment of the PSGL-1 cytoplasmic domain to the actin cytoskeleton is essential for leukocyte rolling on P-selectin. Blood 99:4494–4502

Somers WS, Tang J, Shaw GD, Camphausen RT (2000) Insights into the molecular basis of leukocyte tethering and rolling revealed by structures of P- and E-selectin bound to SLe(X) and PSGL-1. Cell 103:467–479

Sperandio M, Thatte A, Foy D, Ellies LG, Marth JD, Ley K (2001) Severe impairment of leukocyte rolling in venules of core 2 glucosaminyltransferase-deficient mice. Blood 97:3812–3819

Steegmaier M, Blanks JE, Borges E, Vestweber D (1997) P-selectin glycoprotein ligand-1 mediates rolling of mouse bone marrow-derived mast cells on P-selectin but not efficiently on E-selectin. Eur J Immunol 27:1339–1345

Thorpe SJ, Feizi T (1984) Species differences in the expression of carbohydrate differentiation antigens on mammalian blood cells revealed by immunofluorescence with monoclonal antibodies. Biosci Rep 4:673–685

Varki A (1994) Selectin ligands. Proc Natl Acad Sci USA 91:7390–7397

Varki A (1997) Selectin ligands: will the real ones please stand up? J Clin Invest 99:158–162

Vestweber D, Blanks JE (1999) Mechanisms that regulate the function of the selectins and their ligands. Physiol Rev 79:181–213

Weninger W, Ulfman LH, Cheng G, Souchkova N, Quackenbush EJ, Lowe JB, von Andrian UH (2000) Specialized contributions by alpha(1,3)-fucosyltransferase-IV and FucT-VII during leukocyte rolling in dermal microvessels. Immunity 12:665–676

Wild MK, Huang MC, Schulze-Horsel U, van Der Merwe PA, Vestweber D (2001) Affinity, kinetics, and thermodynamics of E-selectin binding to E-selectin ligand-1. J Biol Chem 276:31602–31612

Wilkins PP, McEver RP, Cummings RD (1996) Structures of the O-glycans on P-selectin glycoprotein ligand-1 from HL-60 cells. J Biol Chem 271:18732–18742

Xia L, Ramachandran V, McDaniel JM, Nguyen KN, Cummings RD, McEver RP (2003) N-terminal residues in murine P-selectin glycoprotein ligand-1 required for binding to murine P-selectin. Blood 101:552–559

Xia L, Sperandio M, Yago T, McDaniel JM, Cummings RD, Pearson-White S, Ley K, McEver RP (2002) P-selectin glycoprotein ligand-1-deficient mice have impaired leukocyte tethering to E-selectin under flow. J Clin Invest 109:939–950

Yago T, Leppanen A, Qiu H, Marcus WD, Nollert MU, Zhu C, Cummings RD, McEver RP (2002) Distinct molecular and cellular contributions to stabilizing selectin-mediated rolling under flow. J Cell Biol 158:787–799

Yang J, Galipeau J, Kozak CA, Furie BC, Furie B (1996) Mouse P-selectin glycoprotein ligand-1: Molecular cloning, chromosomal localization, and expression of a functional P-selectin receptor. Blood 87:4176–4186

Yeh JC, Hiraoka N, Petryniak B, Nakayama J, Ellies LG, Rabuka D, Hindsgaul O, Marth JD, Lowe JB, Fukuda M (2001) Novel sulfated lymphocyte homing receptors and their control by a Core1 extension beta 1,3-*N*-acetylglucosaminyltransferase. Cell 105:957–969

9 Saturation Transfer Difference NMR Spectroscopy for Identifying Ligand Epitopes and Binding Specificities

B. Meyer, J. Klein, M. Mayer, R. Meinecke, H. Möller,
A. Neffe, O. Schuster, J. Wülfken, Y. Ding, O. Knaie,
J. Labbe, M.M. Palcic, O. Hindsgaul, B. Wagner, B. Ernst

9.1 Saturation Transfer Difference NMR Principle

We have studied the potential of saturation transfer nuclear magnetic resonance (NMR) experiments in different mode to screen compound mixtures for binding activity and to characterize binding epitopes on the ligand. We have developed a protocol based on the transfer of saturation from the protein to bound ligands which by dissociation is moved into solution where it is detected (Fig. 1; Mayer and Meyer 1999; Peters and Meyer). By subtracting a spec-

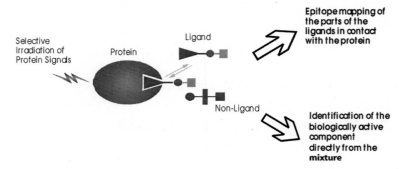

Fig. 1. STD-NMR. When a protein becomes saturated, ligands that are in exchange between the bound and the free form also become saturated when bound to the protein. By chemical exchange that saturation is carried into solution where it is detected. By subtraction of this spectrum from a spectrum without protein irradiation, one obtains an NMR spectrum that has only signals from molecules that bind to the protein. Nonbinders do not show up in the difference spectrum. The receptor protein is saturated with a selective saturation pulse. In general, the saturation pulse consists of a cascade of Gauss-shaped pulses. The duration of saturation times typically ranges from 1 to 2 s. The ligand is normally used in an approximately 50–100-fold molar excess over the protein, allowing one to work with low μM protein concentrations

trum, where the protein is saturated from one without protein saturation, a spectrum is produced where only signals of the ligand(s) remain in the difference spectrum. The irradiation frequency is set to a value where only protein resonances and no resonances of free ligands are located. Usually, irradiation frequencies around 1.5 ppm are practical because no ligand resonances are found in this spectral region, whereas the significant line width of protein signals still allows selective saturation. If the ligands show no resonances in the aromatic spectral region, the saturation frequency may also be placed there.

One major advantage of the new technique is that the saturation transfer difference (STD) protocol may be combined with any NMR pulse sequence, generating a whole suite of STD-NMR experiments such as STD-TOCSY or STD-HSQC (Mayer and Meyer 1999; Peters and Meyer)

The experiment was first used to screen a library of carbohydrate molecules for binding activity towards a carbohydrate binding protein, wheat germ agglutinin (WGA). At the same time it was shown that STD-NMR is useful for determining the binding epitope of the ligand.

In the following, experimental guidelines for STD-NMR are summarized utilizing the original study as an example (Mayer and Meyer 1999). Ligands are added to a solution of the receptor protein and one ^{1}H-NMR experiment is performed where the protein is selectively irradiated at a frequency at least 700 Hz away from the closest ligand signal (on-resonance experiment). Usually, such regions are easily identified depending on the chemical nature of the ligands. Even though the irradiation is highly selective, it yields saturation of the protein by efficient spin diffusion within about 50–200 ms.

A ligand that binds to the protein will be saturated by spin diffusion because it behaves as part of the large protein complex and thus has a reduced mobility in solution. The degree of ligand saturation obviously depends on the residence time of the ligand in the protein binding pocket. The dissociation of the ligand will then transfer this saturation into solution where the free ligand has again narrow line widths. For those ligand protons that interact with protein protons through an intermolecular NOE, a decrease in intensity is observed. However, in the presence of other molecules like impurities and other nonbinding components, it is not usually possible to identify such attenuated signals.

Therefore, in a second experiment the irradiation frequency is set to a value that is far from any signal, ligand, or protein, e.g., 40 ppm (off-resonance spectrum). The spectrum is recorded and yields a normal NMR spectrum of the mixture. Subtraction of the on-resonance from the off-resonance spectra leads to a difference spectrum, in which only signals of protons are visible that were attenuated via saturation transfer. All molecules without binding activity are cancelled out.

A fast off rate of the ligand transfers the information about saturation quickly into solution. If a large excess of the ligand is present, one binding site can be used to saturate many ligand molecules during a time of a few seconds. Ligands in solution lose their information by normal T1 relaxation, which is in the order of about 1 s for small molecules. Thus, the degree of saturated ligands in solution is accumulating during the saturation time. Hereby, the information about the bound state resulting from the saturated protein is amplified. The STD principle is shown in Fig. 1.

On the other hand, if binding is very tight, and consequentially off rates are in the range of several Hz or less, the saturation transfer to ligand molecules is not very efficient. This is usually the case for K_D values below 10^{-10} .

It is obvious that the observed signal intensity of the ligand's signals in the STD-NMR spectrum is not proportional to the binding strength. STD-NMR effects depend largely on the off rate. As outlined above, larger off rates should result in larger STD signals. However, when binding becomes very weak the probability of the ligand being in the receptor site becomes very low and therefore STD-NMR can be used from very tight binding up to a K_D of about 10 mM.

Among other factors, the intensity of the STD signals depends on the irradiation time and on the excess of ligand molecules used. Figure 2 shows the dependence of STD signals as a function of the irradiation time and as a function of the excess of ligand used. The more ligand used and the longer the irradiation time, the stronger the STD signal. Both curves in Fig. 2 asymptotically approach a maximum value. In general, an irradiation time of 2 s and a 50- to 100-fold excess of ligand give good results. The excess of the ligand results in a stronger STD signal – even though a smaller proportion of the ligands become saturated. From the high ligand-to-protein ratios it is clear that the amount of protein required for the measurements is very small. At 500 MHz an amount of approximately 0.1 nmol of protein at µM binding constants is sufficient to record STD spectra. At a molecular weight of 50 kDa, this translates into an approximately 5-µg protein.

Using 1D STD spectra, the compound N-acetylglucosamine (GlcNAc) was identified as the only one with binding affinity for wheat germ agglutinin (WGA). All other molecules showed no response in the STD spectra. As mentioned above, the STD principle can be combined with any NMR pulse sequence, and one rather powerful experiment is the STD-TOCSY experiment. Especially in cases where the library is more complex, the additional deconvolu-

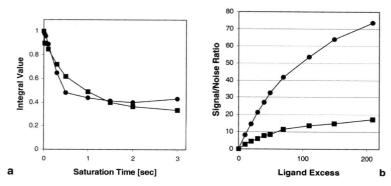

Fig. 2. a Normalized integral values (*I*) of selected ¹H-NMR signals (*circles*, *O*-methyl group of FucOMe; *squares*, H6-methyl group of FucOMe) as a function of saturation time (*t*). The ligand was *O*-methyl-α-L-fucose (FucOMe), the protein was *Aleuria aurantia* agglutinin (AAA). FucOMe was used with 30-fold molar excess over AAA. It is obvious that the decrease of intensity levels out at about 60% of the original intensity. **b** Signal-to-noise ratio (S/N) of STD spectra at 500 MHz as a function of molar excess (*x*) of ligand (GlcNAc; *circles*, *N*-acetyl group; *squares*, H1) over protein (WGA) effect as a function of excess ligand

tion of signals brought about by the second dimension is very helpful. The STD-TOCSY experiment was also shown to be ideally suited for mapping the binding epitope of ligands (Mayer and Meyer 1999, 2001; Haselhorst et al. 2001; Maaheimo et al. 2000).

9.2 STD-NMR for Characterizing Ligand Binding to Membrane Integrated Proteins

In order to study membrane-bound proteins in their native environment, we have studied integrins embedded into liposomes and their binding properties by STD-NMR (Meinecke and Meyer 2001). Many membrane-bound proteins can be stabilized in solution by detergents. The solubilized integrin $\alpha_{IIb}\beta_3$ was integrated into liposome membranes formed of dimyrystyl-phosphatidyl-cholin and dimyrystyl-phosphatidyl-glycerol. These liposomes are known to have a diameter of about 200 nm and carry 50% of the integrins facing to-

wards the inside and 50% facing towards the outside. It is known
that integrins bind to peptides containing the RGD motif. Binding of
such peptides to the liposome-integrated integrins was assayed in
homogenous solution by STD-NMR. It was found that a clear distinction could be made between molecules with binding properties
and other peptides that do not show binding. Also, stronger ligands
could be clearly discriminated from weaker ones by displacing them
from the binding site. We could show that about 0.1 nmol of the
protein is sufficient to assay its binding specificity. The cyclic peptide cyclo(RGDfV) developed by Kessler et al. (Pfaff et al. 1994) to
be a specific inhibitor of integrin $\alpha_V\beta_3$ also has binding affinity to
the integrin $\alpha_{IIb}\beta_3$ with a dissociation constant of 5 µM. This ligand
displaced the open chain peptide RGD from the binding site

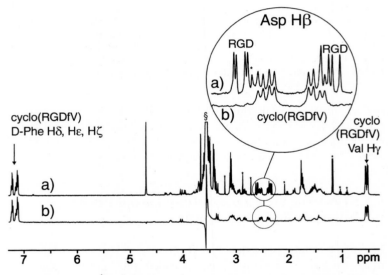

Fig. 3. a Normal ^{1}H-NMR spectrum of 274 µM RGD and 264 µM cy-
clo(RGDfV) with integrin $\alpha_{IIb}\beta_3$ containing liposomes displaying signals of
both ligands. The *inset* shows the expanded region of the two diastereopic
Asp Hβ protons (*§*, TRIS buffer; ***, impurity). **b** STD-NMR spectrum of
274 µM RGD and 264 µM cyclo(RGDfV) with integrin $\alpha_{IIb}\beta_3$ liposomes
showing only STD effects of the tight binding cyclopeptide. The *inset* shows
only signals from the tight binding cyclopeptide

(Fig. 3). It was shown that the peptides have a specific binding to the integrin embedded into liposomes only up to a concentration of about 20 μmol/l. Beyond that, an unspecific binding or potentially a low-affinity binding site on the integrin is used by the peptides. It was shown clearly that no unspecific binding was occurring to the protein-free liposomes.

Further, the binding epitope of the cyclic peptide cyclo(RGDfV) was determined by STD-NMR spectroscopy on liposome integrated integrins. It turned out that the major factor of binding is the phenyl ring of the phenyl alanine, whereas the aspartate has a weaker but still strong contact with the protein, as has the arginine. In the case of the aspartate, the contact is through the carboxylic group, whereas in the case of the arginine it is a hydrophobic contact. The valine was also participating in binding (Fig. 4).

It has been well established that integrins in solubilized form have binding constants to ligands that are about a factor of 100 weaker than the same receptor-ligand interaction in membrane-integrated form. Therefore, it is important to study and understand the interaction of membrane-bound proteins with their ligands in a form close to the native environment.

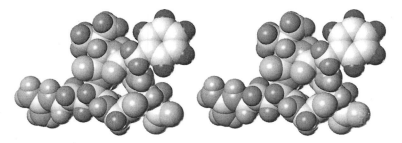

Fig. 4. Binding epitope of cyclo(RGDfV). Combination of the 3D structure of cyclo(RGDfV) in DMSO determined by Aumailley et al. (Aumailley et al 1991; based on NMR spectroscopy and MD simulations) and the STD-NMR derived binding epitopes of the cyclic ligand. The aromatic protons of D-Phe show strong STD effects indicating closest contacts to the integrin. The medium STD intensities of the γ-protons of Val, Arg Hα, Arg Hβ, and Arg Hγ, lead to important binding epitopes. Other section of medium STD intensity are D-Phe Hβ, one β-proton of Asp, and one α-proton of Gly

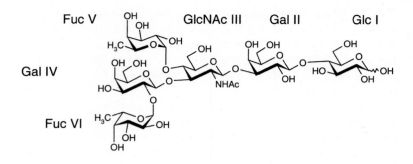

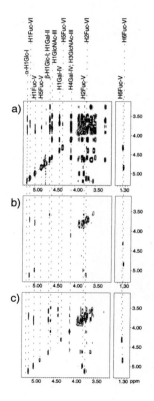

Fig. 5 a–c. Epitope mapping for lacto-*N*-difucosylhexaose I bound to AAA. **a** TOCSY spectrum of ligand only (no protein). **b** STD-TOCSY spectrum in presence of the protein showing only the intensive signals of fucosyl residues V and VI. **c** STD-TOCSY spectrum at about 60% level of that in **b** showing, in addition to the fucosyl cross peaks from **b**, cross peaks originating from galactose IV and *N*-acetylglucosamine III. The full STD-TOCSY exhibiting also the even less intensive signals of galactose II and glucose I (at ≈30% of the intensity of cross peaks in **b** is not shown). The STD-TOCSY spectrum was recorded at 300 K with *on resonance* irradiation at 10 ppm and *off resonance* irradiation at 30 ppm and the irradiation power set to 02 W

9.3 Epitope Mapping with STD-NMR

It was demonstrated that STD-NMR can easily be used to identify the building blocks of ligands in direct contact to the receptor because it receives the highest degree of saturation. The interaction of a hexasaccharide with the lectin *Aleuria aurantia* agglutinin (AAA) shows that only the fucosyl residues directly interact with the protein (Mayer and Meyer 1999; Fig. 5).

9.4 Group Epitope Mapping with STD-NMR

It has been shown (Mayer and Meyer 2001) that epitope mapping can be further refined in order to allow so-called group epitope mapping (GEM). For group epitope mapping it is important that the residence time of the ligand in the bound state is significantly shorter than the T2 time of the ligand in the bound state. We have shown that the groups binding directly to the protein can be identified from STD-NMR spectroscopy (Mayer and Meyer 2001) and that the identification of these groups can be obtained down to the functional group motif. Group epitope mapping can only be observed if the ligand has a fast off rate which is normally the fact for dissociation constants of about $K_D = 0.1$ µM or weaker. Stronger binding normally reduces the off rate so much that the ligand has a too long residence time on the receptor, during which spin diffusion reduces the differences between protons within one residue such that they are insignificant. Thus, the distinction of protons close to the protein surface from those that are not in direct contact with the protein is made virtually impossible. Group epitope mapping has been shown to be possible with peptides, carbohydrates, and aromatic ligands. As an initial example, we used carbohydrate recognition by lectins to identify the functional groups involved in binding to the receptor proteins.

Using the 120-kDa protein *ricinus communis* agglutinin (RCA$_{120}$) as the protein target, the functional groups involved in binding of the ligand methyl β-D-galactoside could be determined. Using STD-NMR experiments, it was shown that only the H2, H3, H4, and H6 protons of the galactose were involved in direct binding to the pro-

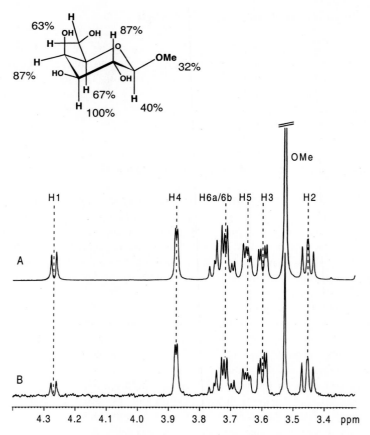

Fig. 6. *Top*: Structure of β-GalOMe and the relative degrees of saturation of the individual protons normalized to that of the H3 proton as determined from 1D STD-NMR spectra at a 100-fold excess: concentration of RCA$_{120}$ was 40 μM and that of β-GalOMe 4 mM. *Bottom*: **A** Reference WATER-GATE NMR spectrum of a mixture of RCA$_{120}$ (40 μM binding sites) and β-GalOMe (12 mM) at a ratio of 1:30. **B** WATERGATE STD-NMR spectrum of the same sample. On resonance irradiation was set to −0.4 ppm over a period of 2 s. Prior to acquisition a 30 ms $T_{1\rho}$ filter was applied to remove residual protein resonances. From the STD spectrum one can characterize the binding epitope by using relative integral intensities of the signals in spectrum B

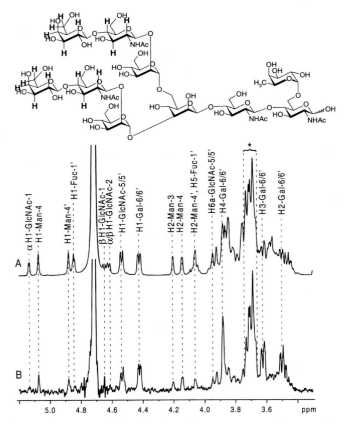

Fig. 7. *Top*: Structure of the ligand: a biantennary complex type decasaccharide (NA$_2$). Highlighted in bold are the groups interacting with RCA$_{120}$ as determined by STD-NMR. *Bottom*: **A** Section of a reference NMR spectrum of a mixture of the RCA$_{120}$ tetramer (50 μM binding sites) and NA$_2$ (0.55 mM) at a ratio of 1:11. **B** STD-NMR spectrum revealing that the directly interacting residues of NA$_2$ have the strongest signals. The resonances corresponding to the terminal galactoses Gal-6/6′ and the adjacent GlcNAc-5/5′ have the most intensive STD signals. The spectral region from 365 to 375 ppm marked with the *asterisk* shows strong STD signals which originate almost entirely from the H5 and H6a/6b of Gal-6/6′ and the H2, H3, and H4 of GlcNAc-5/5′ protons at equal parts (cf. text) H1-Fuc-1′, and α-H1-GlcNAc-1 have almost no detectable STD signal intensity. This shows that they are far away from the binding domain of the lectin

tein (Fig. 6). These data were in perfect agreement with earlier published results obtained from chemical modification of these functional groups (Bhattacharyya and Brewer 1988).

The 1D-spectra of the biantennary decasaccharide at an 11-fold excess over the RCA_{120} show clearly that the terminal residues galactose and *N*-acetyl-glucosamine are most distinctly involved in binding, whereas signals of the sugars close to the reducing end are very low in intensity (Fig. 7). In a more detailed way, the same information can be obtained from the TOCSY spectrum.

Using the large biantennary decasaccharide (Fig. 7) as ligand to RCA_{120}, we could again identify the functional groups involved in binding. In this case, the terminal β-D-galactose residue is involved in binding with its H2, H3, H4, and H6 protons. Furthermore, the H2, H3, and H4 protons of the penultimate β-D-glucosamine residue are also close to the protein surface and contribute to binding. Because of signal overlap, one could not identify the individual share among the protons H2, H3, and H4 with a direct involvement. The other protons of the β-D-GlcNAc residue make a direct contact to the protein. This shows clearly that the glucosamine residue carries a small portion of the binding specificity of that ligand.

The same could be verified for lactosamine as a ligand. It was demonstrated earlier that elongation of a galactose residue by a β-D-glucosamine residue increases binding and that the glucosamine carries about 20% of the binding energy (Sharma et al. 1998).

9.5 Binding Constants from STD-NMR Experiments

The binding constant of ligands can be obtained from STD-NMR spectra. In the example described above, where the biantennary decasaccharide interacts with the receptor RCA_{120}, one can perform a competition titration with the weaker ligand β-D-GalOMe. The results of the titration give STD amplification factors as a function of the concentration of the ligand. The data shown in Fig. 8 clearly prove that the decasaccharide is replaced from the binding side upon adding the weaker methyl-galactoside ligand. The absolute intensity of the β-GalOMe ligand increases with increasing the concentration from 0 to 2.5 measured as the STD amplification factor. The signal intensity of the

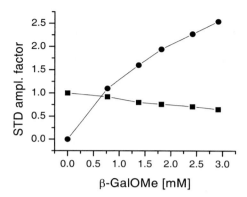

Fig. 8. STD amplification factors (*circles*, H1-β-GalOMe; *squares*, H1-Gal-6/6' NA$_2$) determined from STD spectra on titration of β-GalOMe to a sample of RCA$_{120}$ (50 µM in binding sites) and NA$_2$ (0.55 mM). The STD amplification factor of the signal corresponding to NA$_2$ decreases from 1 to 0.66 with increasing concentration of β-GalOMe. This competition experiment gives evidence to the specificity of the RCA$_{120}$ towards galactose containing saccharides. The K_D of NA$_2$ can be calculated to 27 µM

decasaccharide protons is reduced from 1 to about 0.75 at the same time. This relatively small decrease in STD amplification factor for the decasaccharide compared to the strong increase of the signals for β-GalOMe shows clearly that the latter is a weaker ligand, whereas the former binds stronger to the receptor protein. Using the decrease of the signal intensity of the decasaccharide and knowing the binding constant of the monosaccharide, one can easily determine from a one-site competition model the dissociation constant of the decasaccharide to be $K_D = 27$ µM. Surface plasmon resonance experiments (Biacore) result in a comparable $K_D = 4.4$ µM.

9.6 STD-NMR Analysis of the Binding of a Carbohydrate Library to E-Selectin

In collaboration with Prof. B. Ernst, Basel, Switzerland, and Prof. O. Hindsgaul, Edmonton, Canada, we tried to identify affinity to E-selectin in a randomly sulfated and fucosylated lactose library (Fig. 9).

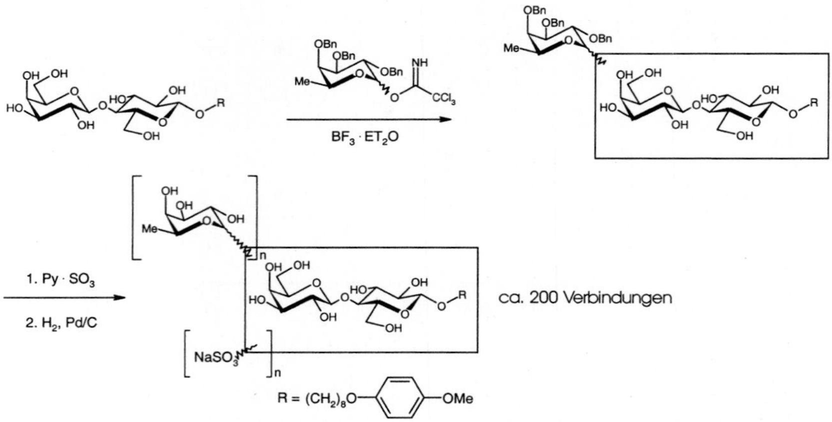

Fig. 9. Synthetic route to the randomly sulfated and randomly fucosylated oligosaccharide library representing sialyl Lewis[X] mimics

They have synthesized a combinatorial library of E-selectin ligands starting from lactose and arriving at a substituted lactose that has randomly fucosyl residues and randomly sulfate residues attached to it. It was unknown where and how many substituents were attached. The reducing end was capped by a spacer. There were about 200 components in the library and a micromolar binding affinity was identified in this library, which is a good K_D for E-selectin. A deconvolution of the library was unsuccessful.

NMR spectra of sugars have an area of large overlap where most of the sugar signals fall within a range of 1 ppm. In the spectra of the library we also have a couple of anomeric signals that are outside the ring protons, and the methyl protons of the fucose are in the high field range. Further, there are some aromatic by products in the sample, because the libraries could not be purified. We used STD spectroscopy to get an idea for binding affinity. TOCSY spectra and saturation transfer difference TOCSY were recorded. In the STD-TOCSY, it is obvious that most signals are eliminated and only very few signals are left, which is clear from an overplot of the expanded regions of the TOCSY and the STD-TOCSY spectrum (cf. Fig. 10). It is obvious that only a few components of this complex library are

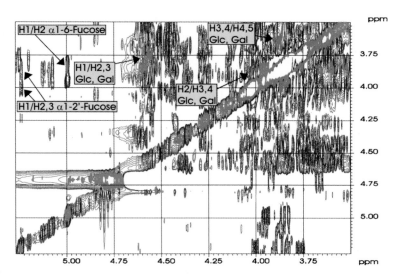

Fig. 10. Expansion from the TOCSY spectra of the oligosaccharide library shown in Fig. 9 and E-selectin as a receptor. *Blue*, TOCSY spectrum of mixture; *red*, STD-TOCSY spectrum showing only a few signals left in the spectrum that arise from 1–3 linked fucosyl lactose moieties and 1–2′ linked fucosyl lactose derivatives. The position of the sulfates could not be assigned unambiguously (cf. text)

responding in the STD-TOCSY spectra. In order to assign the spin systems, selective 1D TOCSY and selective 1D COSY spectra were necessary to get further information about the structures of these molecules. We did not finally succeed in defining the structures, including the location of the sulfate groups. However, we could identify the basic carbohydrate trisaccharide structures that carry binding affinity. It is not surprising that the 3-fucosyl-lactose was identified as one compound with binding affinity. The position of the sulfate could not be located from NMR spectroscopy. It was surprising, however, that also a sulfated 2′-fucosyl-lactose carries binding specificity which is actually, according to STD-NMR, spectra somewhat stronger. Again, we do not know where the sulfate group(s) is (are) located on the trisaccharide. Experimental verification of the results by selectively resynthesizing the sulfated trisaccharides is pending.

From modeling studies it is probable that the 3′ and the 6′ sulfate of the 3-fucosyl-lactose and of the 6-sulfate of the 2′-fucosyl-lactose are potentially potent ligands to E-selectin.

9.7 Boundary Conditions for STD-NMR Spectroscopy

The STD effect can best be viewed if the STD amplification factor (Mayer and Meyer 2001) is being used for the quantification of the response of the ligand in interaction with the receptor protein. The STD amplification factor is obtained by multiplying the percent STD effect of a given proton at a given concentration with the excess of the ligand over the protein. Therefore, the STD amplification factor effectively carries a unit normalized to the intensity of a proton of the protein and is the best measure to assess the sensitivity of the method (cf. Fig. 11).

If one determines the STD effect as a function of saturation time for β-D methyl galactoside in its binding to RCA120, one can clearly see that at an excess of about 12-fold the STD amplification

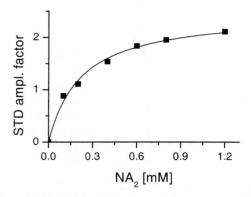

Fig. 11. Titration plot of decasaccharide NA_2 (cf. Fig 7) to NMR sample containing RCA_{120} (20 μM in binding sites) monitoring the increase of the STD amplification factor of the H4-Gal proton versus the ligand concentration (T_{sat}=2 s) I_0=integral of one unsaturated proton, I_{sat}=integral of one proton after saturation

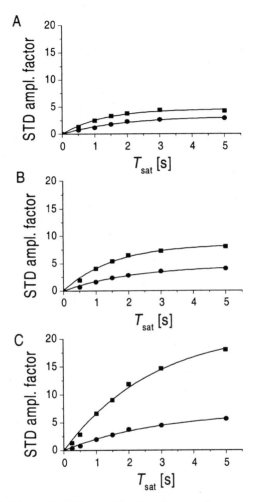

Fig. 12 A–C. Observed STD amplification factors of two resonances of β-GalOMe plotted against the saturation time T_{sat} at three different ligand concentrations (*squares*, H3 proton; *circles*, OMe protons). **A** STD amplification factor at a concentration of 0.5 mM; **B** 1 mM; and **C** 4 mM of β-GalOMe in the presence of 40 µM binding sites of RCA_{120}. A large ligand excess yields larger STD intensities and better discrimination between strongly and weakly binding groups

factor varies between the proton involved in direct binding (H3) and the O-methyl protons by only a factor of about 1.5 (cf. Fig. 12). Both saturation curves have reached their maximum at about 3 s saturation time. Increasing the excess of the ligand (panel B in Fig. 12) to about 25-fold increases the distinction between the proton involved in direct binding (H3 of galactose), and the O-methyl group that is not in close contact with the binding pocket by a factor of more than two. Now even at 5 s saturation time, the maximum of the STD amplification has not yet been reached. Increasing the excess of the ligand further to 100-fold, the absolute magnitude of the response from the ligand increases further (cf. panel C in Fig. 12), and the discrimination between the protons of the O-methyl group and H3 is now better than a factor of three.

The complex type decasaccharide (Fig. 7) has a larger binding constant, and therefore a slower exchange rate. Here, one obtains lower maximal STD amplification factors. The titration curve of the complex type decasaccharide in its binding to RCA120 gives a maximum STD amplification factor of two at 1.2 mM ligand concentration. The reason for the lower STD amplification factor is the decreased off rate (k_{off}) of this ligand associated with the stronger binding.

The examples demonstrate that STD-NMR is well-suited for studying interactions of ligands with macromolecular receptors at a very high sensitivity. All types of NMR spectra can be recorded because STD is a preparatory pulse sequence, e.g., 1D-NMR spectra, TOCSY, COSY, HSQC, and others. STD-NMR can be used to identify from large complex mixtures of compounds those with bioactivity. The largest pool screened contained about 250 compounds. On the receptor side, there is no upper size limit of the protein. However, it should not be smaller than about 10 kDa because the protein has to be effectively saturated by spin diffusion. The receptor proteins can be coupled to solid phase and proteins can also be integrated into liposomes, giving access to screening of membrane-bound or membrane-integrated proteins. The ligands can be either carbohydrates, monosaccharides as well as oligosaccharides up to about 2 kDa, aromatics, peptides, peptidomimetics, and other molecules. The only requirement is that a concentration of the ligand of about 50 µM can be prepared in a solvent system that is also compatible with the receptor protein.

References

Aumailley M, Gurrath M, Müller G, Calvete J, Timpl R, Kessler H (1991) Arg-Gly-Asp constrained within cyclic pentapeptides. Strong and selective inhibitors of cell adhesion to vitronectin and laminin fragment P1. FEBS Lett 291:50–54

Bhattacharyya L, Brewer CF (1988) Lectin-carbohydrate interactions. Studies of the nature of hydrogen bonding between D-galactose and certain D-galactose-specific lectins, and between D-mannose and concanavalin A. Eur J Biochem 176:207–212

Haselhorst T, Weimar T, Peters T (2001) Molecular recognition of sialyl Lewis(x) and related saccharides by two lectins. J Am Chem Soc 123:10705–10714

Maaheimo H, Kosma P, Brade L, Brade H, Peters T (2000) Mapping the binding of synthetic disaccharides representing epitopes of chlamydial lipopolysaccharide to antibodies with NMR. Biochemistry 39:2778–12788

Mayer M, Meyer B (1999) Characterization of ligand binding by saturation transfer difference NMR spectroscopy. Angew Chem 38:1784–1788

Mayer M, Meyer B (2001) Group epitope mapping by saturation transfer difference NMR to identify segments of a ligand in direct contact with a protein receptor. J Am Chem Soc 123:6108–6117

Meinecke R, Meyer B (2001) Determination of the binding specificity of an integral membrane protein by saturation transfer difference NMR: RGD peptide ligands binding to integrin alphaIIbeta3. J Med Chem 44:19–3065

Pfaff M, Tangemann K, Muller B, Gurrath M, Muller G, Kessler H, Timpl R, Engel J (1994) Selective recognition of cyclic RGD peptides of NMR defined conformation by alpha IIb beta 3, alpha V beta 3, and alpha 5 beta 1 integrins. J Biol Chem 269:20233–20238

Sharma S, Bharadwaj S, Surolia A, K Podder S (1998) Evaluation of the stoichiometry and energetics of carbohydrate binding to Ricinus communis agglutinin: a calorimetric study. Biochem J 333:539–542

T Peters, B Meyer, German Pat No 19649359, Swiss Pat No 690695, Verfahren zum Nachweis biologisch aktiver Substanzen in Substanzbibliotheken US Pat No 6,214,561, GB-Patent No GB2321104, Method for Detecting Biologically Active Compounds from Compound Libraries

10   Chemokine Receptor Antagonists from Discovery to the Clinic

R. Horuk

Chemokine receptors belong to one of the most pharmacologically exploited proteins; the G-protein coupled receptors (GPCRs). Drugs that target these receptors make up greater than 45% of all known marketed medicines. The first recorded uses of drugs directed at this important family of proteins can be traced back to ancient Chinese and Indian physicians who were using plant extracts to treat a variety of disorders (Ding 1987; Sevenet 1991). For example, although tetrahydropalmitine, a potent dopamine receptor antagonist, was isolated only a few years ago from the fumewort plant, the plant itself was first described for its tranquilizing effects as early as the fifth century (Ding 1987). Extracts from the deadly nightshade family have been widely used as analgesics and anesthetics in medicine since ancient times (Ding 1987). The active principles, identified as the alkaloids atropine and scopolamine, are potent musacarinic receptor antagonists.

From the ancient shaman who searched for medicinal plants to treat disease (often with much trial and error until the right combinations were found) to the modern pharmaceutical industry with its sophisticated high-throughput mechanism-based screening programs; the quest to find drugs to help the sick and ailing is an ongoing process that has been around since the dawn of mankind. Today the modern pharmaceutical houses concentrate ever-increasing resources and money on finding potent drugs that target both old diseases such as multiple sclerosis and rheumatoid arthritis, and modern diseases such as AIDS and organ transplant rejection. Collectively, the chemokines, because of their important role in these and other diseases, have been the focus of much attention by drug companies, and almost all of the major pharmaceutical houses have screens to identify chemokine receptor antagonists.

Several excellent reviews published recently have concentrated on the biology, pathophysiology and molecular mechanisms of action of the chemokines (Gerard and Rollins 2001; Mackay 2001; Thelen 2001) and the reader is directed toward them to gain a thorough understanding of the importance of this growing family of proteins. Although some background will be given here to aid in an understanding of the medical importance of chemokines, this review will focus on the rapid advances that have been made in identifying and characterizing chemokine receptor antagonists by discussing their efficacy in animal models of disease as well as in detailing their progression through human clinical trials.

10.1 Chemokines and Their Receptors

Chemokines belong to a large family of small, chemotactic cytokines characterized by a distinctive pattern of four conserved cysteine residues (Mackay 2001). They are divided into two major (CXC and CC) and two minor (C and CX3C) groups dependent on the number and spacing of the first two conserved cysteine residues. Although originally identified on the basis of their ability to regulate the trafficking of immune cells, the biological role of chemokines goes well beyond this simple description of their function as chemoattractants, and they have been shown to be involved in a number of biological processes, including growth regulation, hematopoiesis, embryologic development, angiogenesis, and HIV-1 infection (Mackay 2001).

Chemokines mediate their biological effects by binding to cell surface receptors which belong to the GPCR superfamily (Thelen 2001). Receptor binding initiates a cascade of intracellular events mediated by the receptor-associated, heterotrimeric G proteins. These G-protein subunits trigger various effector enzymes, which leads to the activation not only of chemotaxis but also to a wide range of functions in different leukocytes such as an increase in the respiratory burst, degranulation, phagocytosis, and lipid mediator synthesis (Thelen 2001).

Chemokines have been shown to be associated with a number of autoinflammatory diseases including multiple sclerosis, rheumatoid arthritis, atherosclerosis, dermatitis, organ transplant rejection, etc. (Gerard and Rollins 2001). Evidence, reviewed below, is mounting that chemokines may play a major role in the pathophysiology of these diseases and thus chemokine receptor antagonists could prove to be useful therapeutics in treating these and other proinflammatory diseases.

10.2 CC Chemokine Receptor Antagonists

10.2.1 CCR1 Antagonists

Insight into the physiological and pathophysiological roles of CCR1 have been provided by studies with potent CCR1 antagonists (Fig. 1., compound 1; (Hesselgesser et al. 1998; Horuk et al. 2001a, b; Liang et al. 2000) and confirmed by targeted gene disruption studies (Gao et al. 1997, 2000; Gerard et al. 1997; Rottman et al. 2000).

Three separate studies with potent CCR1 receptor antagonists have illuminated the role of CCR1 in the pathophysiology of multiple sclerosis and organ transplant rejection (Horuk et al. 2001a, b;

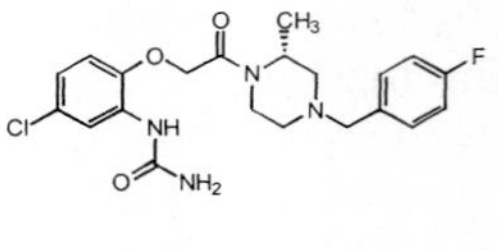

1
Berlex
BX 471
IC_{50} (MIP-1α) = 2 nM
IC_{50} (RANTES) = 6 nM

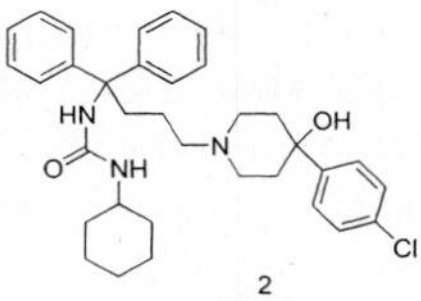

2
Takeda
IC_{50} (MIP-1α) = 8 nM
IC_{50} (RANTES) = 5 nM

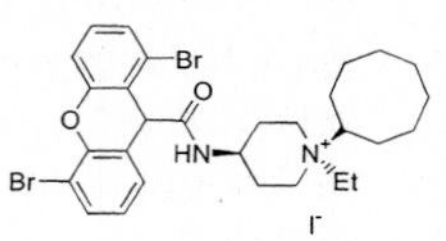

3
Banyu
IC_{50} (CCR1) = 1.9 nM
IC_{50} (CCR3) = 2.7 nM

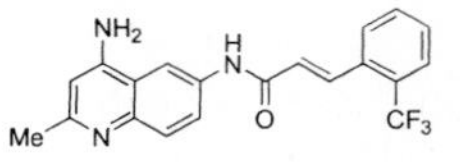

4
Merck
CCR1, CCR2, CCR3, CCR4,
CCR5, CXCR3, CXCR4
no activity reported

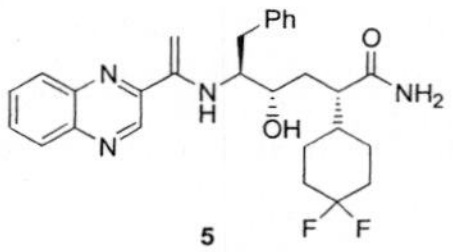

5
Pfizer (CCR1)
no activity reported

Fig. 1. Structures of CCR1 antagonists

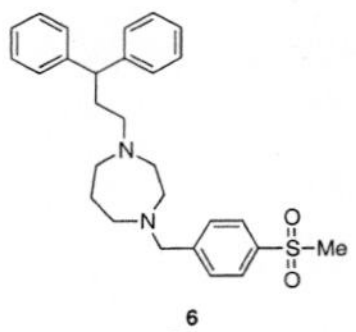

6
Tijian
IC_{50} 9 nM

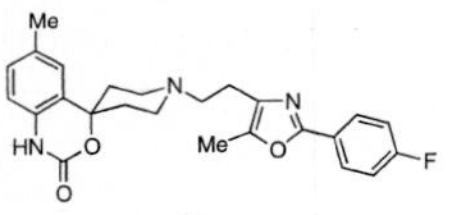

7
Roche Biosciences
IC_{50} 33 nM

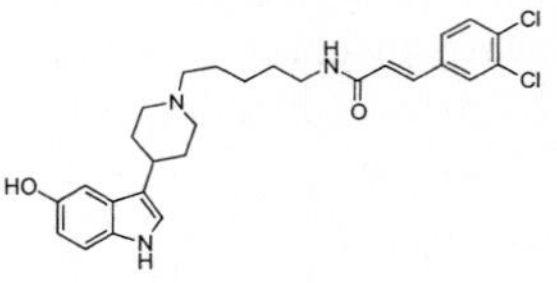

8
SmithKline Beecham
IC_{50} 50 nM

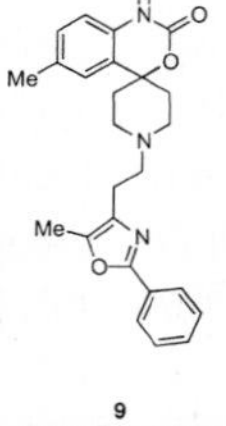

9
Roche Biosciences
IC_{50} 89 nM

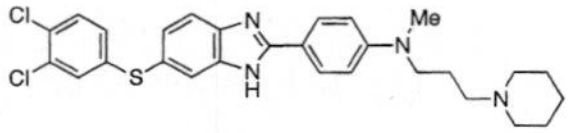

10
LeukoSite/Parke-Davis
IC_{50} 5.5 μM

Fig. 2. Structures of CCR2 antagonists

Liang et al. 2000). Several potent nonpeptide CCR1 antagonists have been reported (Fig. 1); (Hesselgesser et al. 1998; Horuk et al. 2001 a, b; Liang et al. 2000). The most potent member of this class of compounds, BX 471, displaced the CCR1 ligands, MIP-1α, RANTES, and MCP-3, with high affinity and was a potent functional antagonist based on its ability to inhibit a number of CCR1-mediated effects, including Ca^{2+} mobilization, increase in extracellular acidification rate, CD11b expression, and leukocyte migration (Liang et al. 2000). In addition, BX 471 demonstrated a greater than 10,000-fold selectivity for CCR1 compared with 28 different GPCRs. Pharmacokinetic studies demonstrated that BX 471 was orally active with a bioavailability of 60% in dogs.

In a rat experimental autoimmune encephalomyelitis (EAE) model of multiple sclerosis, BX 471 dose-responsively decreased the clinical score. At the highest dose of 50 mg/kg, BX 471 reduced the clinical score by around 50% (Liang et al. 2000). The much higher doses of BX 471 that are required to be effective in rat EAE are due to the fact that the compound has an IC_{50} of 121 nM for inhibition of MIP-1α binding to rat CCR1, compared with an IC_{50} of 1–2 nM for human CCR1. Based on these considerations, it is likely that much lower doses of BX 471 (500 µg/kg or less) would be required to be therapeutically effective in treating multiple sclerosis in humans.

The CCR1 receptor antagonist BX 471 is also efficacious in a rat heterotopic heart transplant rejection model (Horuk et al. 2001 a). Treatment of animals with BX 471 and a subtherapeutic dose of cyclosporin, 2.5 mg/kg, which is by itself ineffective in prolonging transplant rejection, was much more efficacious in prolonging transplantation rejection than treatment of animals with either cyclosporin or BX 471 alone. Immunohistology of the rat hearts for infiltrating monocytes confirmed these data. Three days after transplantation, the extent of monocytic graft infiltration was significantly reduced by the combined therapy of BX 471 and cyclosporin. Thus, BX 471 given in combination with cyclosporin resulted in a clear increase in efficacy in heart transplantation compared to cyclosporin alone. These data were in line with the observed effects of BX 471 in dose responsively blocking the firm adhesion of monocytes triggered by RANTES on inflamed endothelium. Together, these data demonstrate a significant role for CCR1 in allograft rejection.

Two recent studies with CCR1(–/–) mice have confirmed the roles of CCR1 in the pathophysiology of multiple sclerosis and organ transplant rejection (Gao et al. 2000; Rottman et al. 2000). In the first study, Rottman et al. (Rottman et al. 2000) demonstrated, in an EAE model of multiple sclerosis, that CCR (–/–) mice had a significantly reduced incidence of disease compared to wild-type mice. The spinal cords of the wild-type mice showed nonsuppurative myelitis, while those from the CCR1 knockouts were minimally inflamed. Taken together with the CCR1 antagonist studies discussed above (Liang et al. 2000), these data strongly argue that CCR1 plays a role in the pathogenesis of EAE and further suggest a role for CCR1 in the pathophysiology of the human disease, multiple sclerosis. In the second study, Gao et al (Gao et al. 2000) reported a significant prolongation of allograft survival in CCR1(–/–) mice in four separate models of cardiac allograft rejection. In one model, levels of cyclosporin that had marginal effects in CCR1(+/+) mice resulted in permanent allograft acceptance in CCR1(–/–) recipients. These studies and those described above with the CCR1 receptor antagonist (Horuk et al. 2001a) suggest that therapies to inhibit CCR1 may prove useful in preventing acute and chronic rejection clinically.

A number of other companies including Takeda, Banyu, Merck, and Pfizer have also disclosed CCR1 inhibitors (Fig. 1, compounds 2–5; a more complete account can be found in a recent review Horuk and Ng 2000).

10.2.2 CCR2 Antagonists

CCR2 is mainly expressed in monocytes and responds to a number of CC chemokines, including MCP-1, MCP-2, MCP-3, and MCP-4. A number of studies indicate that CCR2 may play a role in the pathogenesis of multiple sclerosis (Fife et al. 2000; Izikson et al. 2000) and in atherogenesis (Boring et al. 1998; Dawson et al. 1999) and for these reasons the receptor has attracted some attention as a therapeutic target.

A number of small molecule CCR2 antagonists have been described (Fig. 2). Most of the compounds known to selectively inhibit binding to CCR2 act by antagonism of its ligand, MCP-1. One such compound was reported by Teijin (Shiota et al. 1997; Fig. 2, compound 6) and shares many of the structural features of the CCR1 antagonists discussed above. It possesses two phenyl groups linked to a substituted heterocyclic ring by an alkyl chain. It had a reported activity of 9 μM in a chemotaxis assay and did not affect MIP-1α mediated chemotaxis, implying functional selectivity for inhibition of binding to CCR2 over CCR1. No in vivo data have been reported.

A more potent antagonist of MCP-1 has been reported by Roche Biosciences (Fig. 2, compound 7) to have an activity of 33 nM (Lapierre et al. 1998). However, it was shown to be weakly effective in inhibiting in vitro monocyte chemotaxis with an IC_{50} of 1 μM. No in vivo data have been reported for this compound.

In a recent communication Forbes et al (Forbes et al. 2000) reported the conversion of a weak CCR2 antagonist, a biphenylated indole piperidine with an IC_{50} of 5.3 μM, to a more potent lead compound. The optimization, which included substitution of the indole ring, replacement of the biphenyl group by a dichloro benzyl group, and introduction of a cinnamide linker resulted in a close to 100-fold increase in binding activity with an IC_{50} of 50 nM (Fig. 2, compound 8). The more potent compound was a true antagonist inhibiting MCP-1-mediated chemotaxis with an IC_{50} of 25 nM. The antagonist was specific for CCR2 over other chemokine receptors tested and its affinity for CCR5 for example was 4260 nM. However, during the course of optimization the compound picked up an unwanted specificity for the 5-HT receptor.

An interesting group of CCR2 antagonists, the spiropiperidines, have been used to model the CCR2 receptor binding site (Mirzadegan et al. 2000). One of the more potent compounds from this group blocks MCP-1 binding to CCR2 with an IC_{50} of 89 nM, and demonstrates specificity since it does not inhibit binding of CXCR1, CCR1, or CCR3 (Fig. 2, compound 9). The molecular basis for this antagonism of CCR2 appears to involve an ionic interaction of the basic nitrogen of the spiropiperidine with an acidic glutamate at position 291 (Glu 291) in the sequence.

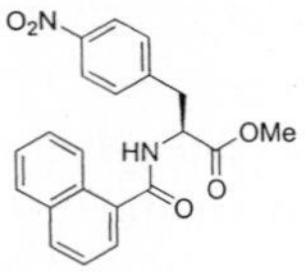

11
SmithKline Beecham
IC_{50} binding = 5 nM
IC_{50} chemotaxis = 25 - 55 nM

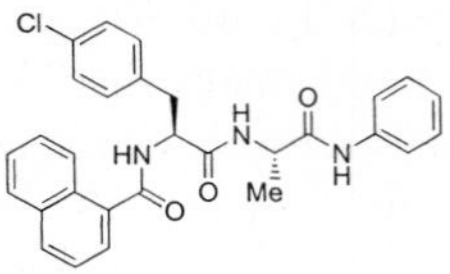

12
SmithKline Beecham
IC_{50} binding = 5 nM
IC_{50} chemotaxis = 15 nM

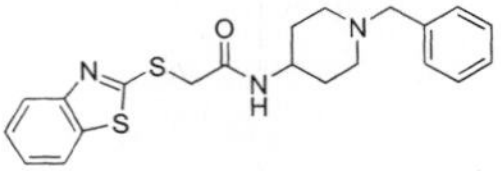

13
Banyu
IC_{50} binding CCR3 = 750 nM
IC_{50} binding CCR1 = 7200 nM

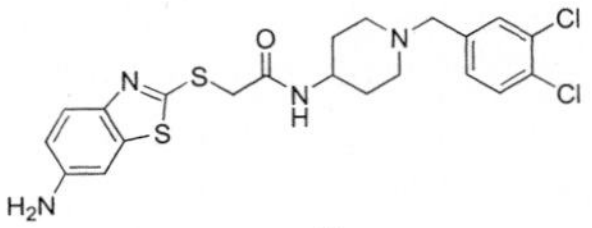

14
Banyu
IC_{50} binding CCR3 = 2.3 nM
IC_{50} binding CCR1 = 1900 nM

Fig. 3. Structures of CCR3 antagonists

10.2.3 CCR3 Antagonists

CCR3 is primarily found on eosinophils and in a subset of Th2 T-cells (Combadiere et al. 1995; Heath et al. 1997). A variety of evidence, including CCR3 antibody blocking studies, has implicated CCR3 in the regulation of eosinophil migration (Heath et al. 1997). These results demonstrate the importance of CCR3 for eosinophil responses and suggest the feasibility of antagonizing this receptor. Based on this data and tissue expression of CCR3 and its ligands from pathophysiological samples (Ying et al. 1999), it is likely that CCR3 is more involved in Th2-mediated responses and could play an important role in allergy research, including asthma and atopic dermatitis. Two separate groups have described targeted gene disruption studies of the CCR3 ligand eotaxin (Rothenberg et al. 1997; Yang et al. 1998) and one group has described CCR3 knockouts (Gerard and Rollins 2001). The eotaxin-deficient mice developed normally and had no histologic or hematopoietic abnormalities

(Rothenberg et al. 1997; Yang et al. 1998). However, while one group demonstrated that eotaxin deficient mice had impaired eosinophil recruitment (Rothenberg et al. 1997), the other group showed that the lack of eotaxin had no effect on the recruitment of eosinophils in a variety of animal models (Yang et al. 1998). Based on these data it is hard to conclude what the role of CCR3 in asthma might be, and data from CCR3 knockout animals further compounds this confusion (Gerard and Rollins 2001). In these studies, reported by Gerard (Gerard and Rollins 2001), it appears that even though CCR3 disruption, in an ovalbumin-induced model of airway hyperactivity, reduced airway eosinophil accumulation by 50%, the animals were not protected. In fact, CCR3$^{-/-}$ mice actually showed a 50% enhancement of bronchial constriction in response to methacholine.

Clarification of the role of CCR3 in the pathophysiology of asthma will probably be aided by the availability of CCR3 antagonists and several have now been described. In a recent communication, the discovery and initial structure–activity relationships of a series of highly selective and potent phenylalanine-derived CCR3 antagonists were described (Dhanak et al. 2001 a, b). High-throughput screening by inhibiting eotaxin-induced intracellular calcium mobilization of RBL-2H3 cells expressing human CCR3 was used to identify antagonists. Chemical optimization of an *N*-benzoyl-3,5-diiodotyrosine ethyl ester using a solution-based parallel synthesis approach identified a compound (Fig. 3, compound 11) that inhibited CCR3 binding with an IC$_{50}$ of 5 nM, and inhibited eotaxin, eotaxin-2, and MCP-4-induced eosinophil chemotaxis at a concentration of 25–55 nM. Furthermore the compound showed greater than 2500-fold selectivity for CCR3 compared to a panel of GPCRs. In order to overcome the presence of a metabolically labile ester bond, which would probably limit its in vivo activity, compound 20 was further optimized and a 4-chlorophenylanine derivative (Fig. 3, compound 12) was identified which was comparable in activity (IC$_{50}$ of binding inhibition of 5 nM, and inhibition of eotaxin-induced eosinophil chemotaxis at a concentration of 15 nM) and metabolically stable.

Screening by inhibition of ^{125}I-eotaxin identified a 2-(benzothiazolethio)acetamide derivative that was a dual CCR1 and CCR3 antagonist (Fig. 3, compound 13; Naya et al. 2001). Optimization of

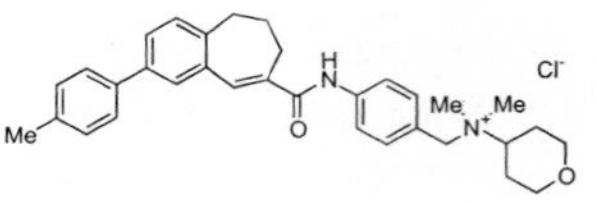

15
Schering Plough
IC$_{50}$ (RANTES) = 2 nM
IC$_{50}$ inhibition viral replication = 6.5 nM

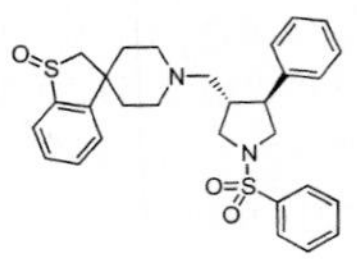

16
Merck
IC$_{50}$ (MIP-1α) = 26 nM

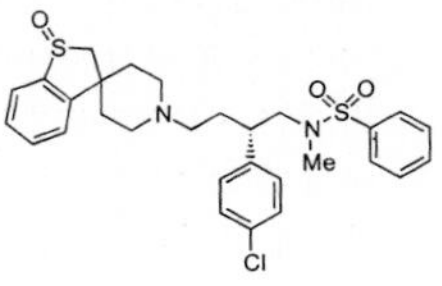

17
Merck
IC$_{50}$ (MIP-1α) = 10 nM
IC$_{90}$ viral inhibition = 1500 nM

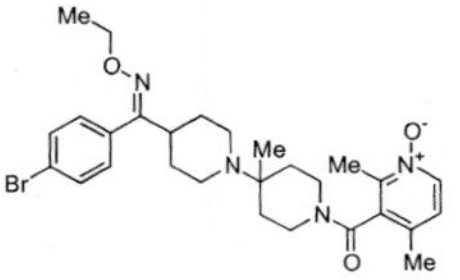

18
Takeda
IC$_{50}$ (RANTES) = 1.4 nM

Fig. 4. Structures of CCR5 antagonists

this early lead by incorporating substituents into each benzene ring of the benzothiazole and piperidine side chains resulted in the discovery of a compound (Fig. 3, compound 14) that exhibited over 800-fold selectivity for CCR3 (IC$_{50}$=2.3 nM) over CCR1 (IC$_{50}$=1,900 nM). This compound also showed potent functional antagonist activity for inhibiting eotaxin (IC$_{50}$=27 nM) or RANTES (IC$_{50}$=13 nM) induced Ca^{2+} increases in eosinophils.

10.2.4 CCR5 Antagonists

The unmasking several years ago of the chemokine receptors CCR5 and CXCR4 as major coreceptors, along with CD4, for human immunodeficiency virus (HIV-1) invasion provided a strong impetus for the rapid development of chemokine receptor antagonists by the pharmaceutical industry. The early stages of HIV-1 infection appear to involve macrophage-tropic strains of HIV-1 known as R5, which use mainly CCR5 as coreceptors (D'Souza and Harden 1996).

HIV-1 resistance exhibited by some exposed but uninfected individuals (Paxton et al. 1996) is due, in part, to a 32-bp deletion in the CCR5 gene (CCR5Δ32), which results in a truncated protein that is not expressed on the cell surface (Liu et al. 1996; Samson et al. 1996). About 1% of Caucasians are homozygous for the CCR5Δ32 allele and appear to be healthy with no untoward signs of disease (Liu et al. 1996; Samson et al. 1996). In fact, recent findings suggest that homozygosity for the CCR5Δ32 alleles confers other selective advantages to these individuals, rendering them less susceptible to rheumatoid arthritis (Garred et al. 1998) and asthma (Hall et al. 1999) and prolonging survival of transplanted solid organs (Fischer-eder et al. 2001).

The clinical relevance of CCR5 in solid organ transplantation was assessed by looking at the effect of the CCR5Δ32 alleles in renal transplant survival. This study examined a total of 1,227 renal-transplant recipients, 21 of whom were homozygous for CCR5Δ32. Analysis of the data demonstrated that individuals who were homozygous for CCRΔ32 had a survival advantage over individuals homozygous for wild-type CCR5. In fact, only one of the 21 CCR5Δ32 patients lost transplant function during follow-up, compared with 78 of the 555 patients with a CCR5 wild-type or heterozygous CCRΔ32 genotype. This study shows that CCR5 can play an important role in enhancing long-term allograft survival, and together with the studies referenced above, underscore the fact that CCR5 antagonists could be therapeutically useful in a variety of clinical situations including organ transplantation, asthma, rheumatoid arthritis, and HIV-1 infection.

A number of pharmaceutical companies, including Schering Plough, SmithKline Beecham, Pfizer, Millennium, Merck, and Takeda (Fig. 4) have programs aimed at identifying CCR5 antagonists. There is limited information available on many of these programs; however, Schering Plough appears to be in phase I clinical trials with their antagonist (Fig. 4, compound 15), and an update on the development of their small molecule CCR5 antagonist, SCH C, was recently given at a clinical meeting (Reyes 2001). This small molecule is a piperidinyl piperidine derivative discovered by high

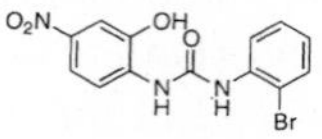

19
SmithKline Beecham
CXCR2: IC_{50} = 22 nM

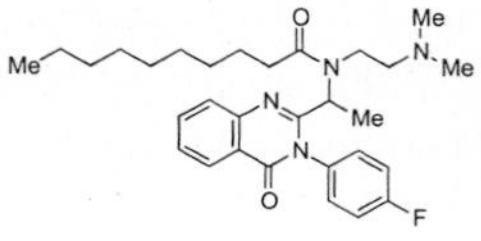

20
ChemoCentryx
CXCR3: IC_{50} = 800 nM

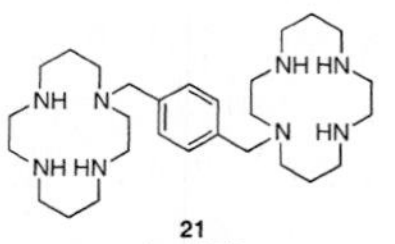

21
AnorMED
CXCR4: IC_{50} = 74 nM
HIV-1: IC_{50} in vitro = 1-10 nM

Fig. 5. Structures of CXC chemokine receptor antagonists

throughput screening in CCR5 binding assays. SCH-C showed specificity for CCR5, inhibiting RANTES binding with an IC_{50} of 2 nM, and has no effect on binding to other chemokine receptors including CXCR4. The compound is a full CCR5 antagonist and can inhibit HIV replication in CCR5-expressing cell lines as well as in a number of human HIV isolates. It inhibits viral entry and replication with IC_{50}s of 0.69 nM and 0.06–6.5 nM, respectively. In addition, it appears to work synergistically with other antiretroviral compounds, such as zidovudine and indinavir.

In a SCID mouse model of HIV infection, SCH C was able to effectively reduce viral replication. SHC C has a good pharmacokinetic profile and single dosing appears to give protection against viral infection for up to 24 h. It works well in inhibiting a number of HIV isolates as well as acting against antiretroviral resistant HIV strains. The drug is orally bioavailable (60%–90%) in a number of species, including man, has low protein binding, and appears not to induce metabolic enzymes. In single-dose phase I safety studies in normal human volunteers at doses ranging from 25 to 600 mg, dose proportional increases in the area under the curve (AUC) were seen, that achieved drug concentrations that were well above the IC_{90} concentration required for the inhibition of most viruses. The drug appeared to be well tolerated; however, some QTc prolongation was seen at the higher doses that could increase the risk of arrhythmias.

Merck recently described a number of potent CCR5 inhibitors discovered in high throughput screening assays that were based on their ability to displace ^{125}I-MIP-1α from CHO cell membranes expressing CCR5. The most recent communication described a series of 1,3,4-trisubstituted pyrrolidines, and the lead compound from this series (Fig. 4, compound 16) had an IC_{50} of 26 nM that showed no agonist activity measured by microphysiometry. The ability of the pyrrolidine compounds to displace ^{125}I-MIP-1α from the CCR5 receptor was shown to be dependent on the stereochemistry of the pyrrolidine scaffold. Although these compounds were potent CCR5 inhibitors, there was no discussion regarding their potential to inhibit HIV. In contrast, in two further communications Merck described a series of piperidinyl butanes and demonstrated that some of these compounds were also able to inhibit viral replication in an in vitro assay. The most potent CCR5 inhibitor from this series (Fig. 4, compound 17) had an IC_{50} of 10 nM in the chemokine binding assay and an IC_{90} of viral inhibition of 1500 nM against the HIV-1 strain, YU-2. In addition, the lead compound showed specificity for CCR5 since it had IC_{50}s greater than 10 µM for inhibition of binding for CCR1, CCR2, CCR3, and CXCR4. Further demonstration of specificity was provided by the finding that the lead compound had no antiviral activity with the X4-tropic NL4–3 strain, which utilizes CXCR4 as the major coreceptor. Finally, pharmacokinetic data indicated that this compound had a short half-life in the rat, 0.7 h, and an oral availability of only 3%.

In addition to the CCR5 antagonists described above, a recent report by Takeda describes a potent nonpeptide inhibitor of CCR5, TAK-779, that blocks HIV-1 infection (Fig. 4, compound 18; Baba et al. 1999). In this study, TAK-779 antagonized the binding of RANTES to CCR5-expressing CHO cells (IC_{50} of 1.4 nM) and blocked CCR5-mediated Ca^{2+} signaling at nanomolar concentrations. Although the compound did not affect binding to CCR1, CCR3, or CCR4, it did show a modest inhibition of binding to CCR2 (IC_{50} of 27 nM) thus it was not truly CCR5 selective.

10.3 CXC Chemokine Receptor Antagonists

10.3.1 CXCR2 Antagonists

CXCR2, together with CXCR1, is the major interleukin (IL-)8-responsive receptor on neutrophils. CXCR2 has also been identified on T cells, monocytes, melanoma cells, endothelial cells, synovial fibroblasts, and Purkinje cells in the brain but so far its function in these cells has not been determined (Murphy et al. 2000). The receptor binds a large number of CXC chemokines that express the ELR motif (Hébert et al. 1991), including IL-8 and MGSA. There is abundant evidence to support a pathological role for CCR2 in a number of inflammatory diseases, including adult respiratory distress syndrome, chronic airway disease, and psoriasis. In addition, it is known that IL-8 and a number of other ELR-containing CXCR2 ligands are strongly angiogenic and their presence in a variety of tumors has aroused speculation that CXCR2 could play an important role in tumor angiogenesis and neovascularization (Strieter et al. 1995).

A potent nonpeptide CXCR2 antagonist, SB 225002 (N-(2-hydroxy-4 nitrophenyl)-N'-(2-bromophenyl)urea), has been reported by SmithKline Beecham (Fig. 5, compound 19; White et al. 1998). It was shown to be selective for inhibition of CXCR2 over CXCR1. SB 225002 was shown to be over 150-fold more potent for inhibition of CXCR2 than CXCR1, or than the receptors for fMLP, LTB$_4$, and LTD. In vitro efficacy was demonstrated by dose-responsive inhibition of calcium mobilization with IC$_{50}$ values of 8–10 nM. In addition, inhibition of neutrophil chemotaxis using human PMNs showed an IC$_{50}$ value of 20–60 nM. These researchers then demonstrated similar activity for inhibiting rabbit neutrophil chemotaxis (30–70 nM) and followed that up with in vivo studies. In a rabbit model of neutrophil activation and attachment to endothelial lung cells, SB 225002 was shown to dose dependently inhibit this effect when IL-8 was administered, but not when fMLP was administered, thereby demonstrating both efficacy and selectivity in vivo.

10.3.2 CXCR3 Antagonists

CXCR3 is highly expressed in IL-2-activated T lymphocytes, but is not detectable in resting T lymphocytes, B lymphocytes, monocytes, or granulocytes. It mediates Ca^{2+} mobilization and chemotaxis in response to ITAC, IP10, and MIG. These observations suggest that CXCR3 is involved in the selective recruitment of effector T cells (Murphy et al. 2000).

Recently studies with CXCR3-deficient (CXCR3(–/–) mice showed a significant prolongation of allograft survival in three in vivo models, demonstrating a role for CXCR3 in the development of transplant rejection (Hancock et al. 2000). First, CXCR3(–/–) mice showed profound resistance to development of acute allograft rejection. Second, CXCR3(–/–) allograft recipients treated with a brief, subtherapeutic course of cyclosporin A maintained their allografts permanently and without evidence of chronic rejection. Third, CXCR(+/+) mice treated with an anti-CXCR3 monoclonal antibody showed prolongation of allograft survival, even if begun after the onset of rejection. These data strongly suggest that CXCR3 plays a key role in T cell activation, recruitment, and allograft destruction.

There is only one report in the patent literature of a CXCR3 antagonist (Schall et al. 2001). Compounds based on a dihdroquinazoline scaffold are cited and a structure (Fig. 5, compound 20) with an affinity of 800 nM is disclosed. The stated compound is claimed to be specific and not to crossreact with five other chemokine receptors. No further information is given.

10.3.3 CXCR4 Antagonists

As discussed for CCR5 inhibitors, the effort expended by the major pharmaceutical companies in discovering potent CXCR4 inhibitors has been fueled by the finding that CXCR4 is a major coreceptor for HIV-1 infection. However, as we shall see later, one of the benefits of this AIDS-driven effort aimed at finding potent CXCR4 antagonists with which to treat this lethal disease, is that the drugs discovered from this approach could also be of potential benefit in treating other diseases in which CXCR4 has recently been postulated to play a role (Muller et al. 2001).

Two peptide antagonists of CXCR4 have been described (Doranz et al. 1997; Murakami et al. 1997). The first is a potent 18-residue oligopeptide antagonist known as T22, and the second a highly cationic oligopeptide containing nine Arg residues known as ALX40-4C. Both of these molecules specifically inhibit the ability of T cell-tropic strains of HIV-1, which use CXCR4, but not M-tropic strains, which utilize CCR5 (Doranz et al. 1997; Murakami et al. 1997). Because of limitations with oral availability, however, these peptides are not as attractive therapeutically as small molecule antagonists.

In addition to the peptide inhibitors, a small molecule heterocyclic bicyclam compound (Fig. 5, compound 21), AMD-3100, previously known to block HIV-1 replication in the nanomolar concentration range (Schols et al. 1997), was shown to inhibit the binding of SDF-1 and 12G5 to CXCR4, and to neutralize CXCR-4-dependent virus infection. AMD-3100 is a specific CXCR4 inhibitor and does not inhibit the binding of CC chemokine ligands to CCR1, CCR2, or CCR5. Further evidence of its specificity is demonstrated by the fact that it is active against T-tropic HIV strains which use CXCR4 as coreceptor, but inactive against M-tropic virus strains which use CCR5 as coreceptor.

AMD3100 is a true antagonist and at 100 ng/ml completely blocks SDF-1-induced Ca^{2+} transients in both SUP-T1 and THP-1, but has no effect on Ca^{2+} flux induced by RANTES, MIP-1a, and MCP-3 in THP-1 cells. AMD-3100 has no agonist properties measured by Ca^{2+} flux studies. Recently, phase I clinical trials with AMD-3100 were described (Hendrix et al. 2000). Single dose escalation studies in normal healthy volunteers by intravenous infusion or by subcutaneous injection were extremely well tolerated at drug doses up to 80 μg/kg. Multiple dosing protocols in which subjects received two escalating oral doses of 80 and 160 μg/kg were also well tolerated without any apparent toxicity. However, the oral bioavailability of the drug was low, less than 2%, and in line with that previously measured in rats, 3.9%. All subjects experienced a dose-related elevation of the white blood cell count, from 1.5 to 3.1 times the baseline, which returned to baseline 24 h after dosing. AMD-3100 demonstrated dose proportionality for the maximum drug concentration in serum (C_{max}) and the area under the concentration-time curve from 0 h to (AUC_{0-}) over the entire dose range. At a single i.v. dose of 80 µg/kg, the concentrations of AMD-3100 stayed well above the in vitro antiretroviral 90% inhibitory concentrations for 12 h. Based on the positive response from these phase I studies, AMD-3100 has recently entered phase II clinical trials in HIV-infected individuals.

An interesting approach recently described is the synthesis of a new drug entity made by covalently linking AMD-3100 to AZT (Dessolin et al. 1999). A number of different constructs were prepared and one, compound 28, showed almost comparable activity to AMD-3100 in inhibiting CXCR4 binding. This prodrug was also quite stable in plasma and had a half-life of around 8 h. The proposed therapeutic advantage of this drug chimera is to create in a single drug entity the ability to inhibit two crucial steps in HIV infection, fusion, and reverse transcriptase activity.

The pathophysiological role of CXCR4 has recently been expanded by the finding that it appears to be highly expressed in a number of human breast cancer cells, malignant breast tumors, and metastases (Muller et al. 2001). This expression is complemented by the fact that SDF-1 expression is the highest in lymph node, lung, liver, and bone marrow, distributions that exactly correspond to regions where one would expect to find breast cancer metastasis appearing. These findings point to an important role for CXCR4 in oncology. Data demonstrating that breast cancer cells signaling through CXCR4 can mediate activities such as actin polymerization and pseudopodia formation, functions that would support the ability of these cells to migrate and become invasive, further supports this. In addition, a neutralizing antibody to CXCR4 significantly impaired the metastasis of breast cancer cells to regional lymph nodes and lung in an animal model. These studies demonstrate that the chemoattractant properties of the CXCR4/SDF-1 interaction, which are so important during morphogenesis in helping to keep cells together to form blood vessels (Tachibana et al. 1998), can also have a darker side in helping to propagate and promote tumor growth. Finally, they highlight another therapeutic opportunity in which CXCR4 small molecule antagonists could be useful.

10.4 Conclusion

In the late 1980s scientists isolated the signaling molecules, chemokines, that allowed leukocytes to communicate with one another and seek out and destroy invading pathogens. However, the immune response is a double-edged sword and can under certain circumstances be inappropriately activated and targeted toward normal healthy tissue, leading to autoimmunity and disease. It was soon realized that an understanding of the mechanisms involved in these processes could provide a key for the identification of successful therapeutic approaches to treat these diseases. The identification of chemokine receptors as a subfamily of GPCRs has paved the way toward the realization of these early goals. In the space of four short years numerous nonpeptide chemokine receptor antagonists have been identified and some are even in clinical trials in man. As the role of chemokine receptors has expanded from autoimmunity to AIDS, the importance of these intervention therapies has grown with them. The promise of highly specific therapies for a number of devastating diseases is on the horizon thanks to the identification of chemokine receptor antagonists, and we can look forward with anticipation to the day when these drugs are finally marketed as potent therapeutics.

References

Baba M, Nishimura O, Kanzaki N, Okamoto M, Sawada H, Iizawa Y, Shiraishi M, Aramaki Y, Okonogi K, Ogawa Y, Meguro K, Fujino M (1999) A small-molecule, nonpeptide CCR5 antagonist with highly potent and selective anti-HIV-1 activity. Proc Natl Acad Sci U S A 96:5698–5703

Boring L, Gosling J, Cleary M, Charo IF (1998) Decreased lesion formation in $CCR2^{-/-}$ mice reveals a role for chemokines in the initiation of atherosclerosis. Nature 394:894–897

Combadiere C, Ahuja SK, Murphy PM (1995) Cloning and functional expression of a human eosinophil CC chemokine receptor. J Biol Chem 270:16491–16494

Dawson TC, Kuziel WA, Osahar TA, Maeda N (1999) Absence of CC chemokine receptor-2 reduces atherosclerosis in apolipoprotein E-deficient mice. Atherosclerosis 143:205–211

Dessolin J, Galea P, Vlieghe P, Chermann JC, Kraus JL (1999) New bicyclam-AZT conjugates: design, synthesis, anti-HIV evaluation, and their interaction with CXCR-4 coreceptor. J Med Chem 42:229–241

Dhanak D, Christmann LT, Darcy MG, Jurewicz AJ, Keenan RM, Lee J, Sarau HM, Widdowson KL, White JR (2001a) Discovery of potent and selective phenylalanine-derived CCR3 antagonists. Part 1. Bioorg Med Chem Lett 11:1441–1444

Dhanak D, Christmann LT, Darcy MG, Keenan RM, Knight SD, Lee J, Ridgers LH, Sarau HM, Shah DH, White JR, Zhang L (2001b) Discovery of potent and selective phenylalanine derived CCR3 receptor antagonists. Part 2. Bioorg Med Chem Lett 11:1445–1450

Ding GS (1987) Important Chinese herbal remedies. Clin Ther 9:345–357

Doranz BJ, Grovit-Ferbas K, Sharron MP, Mao SH, Goetz MB, Daar ES, Doms RW, O'Brien WA (1997) A small-molecule inhibitor directed against the chemokine receptor CXCR4 prevents its use as an HIV-1 coreceptor. J Exp Med 186:1395–1400

D'Souza MP, Harden VA (1996) Chemokines and HIV-1 second receptors – Confluence of two fields generates optimism in AIDS research. Nature Med 2:1293–1300

Fife BT, Huffnagle GB, Kuziel WA, Karpus WJ (2000) CC chemokine receptor 2 is critical for induction of experimental autoimmune encephalomyelitis. J Exp Med 192:899–906

Fischereder M, Luckow B, Hocher B, Wuthrich RP, Rothenpieler U, Schneeberger H, Panzer U, Stahl RA, Hauser IA, Budde K, Neumayer H, Kramer BK, Land W, Schlondorff D (2001) CC chemokine receptor 5 and renal-transplant survival. Lancet 357:1758–1761

Forbes IT, Cooper DG, Dodds EK, Hickey DM, Ife RJ, Meeson M, Stockley M, Berkhout TA, Gohil J, Groot PH, Moores K (2000) CCR2B receptor antagonists: conversion of a weak HTS hit to a potent lead compound. Bioorg Med Chem Lett 10:1803–1806

Gao JL, Wynn TA, Chang Y, Lee EJ, Broxmeyer HE, Cooper S, Tiffany HL, Westphal H, Kwon-Chung J, Murphy PM (1997) Impaired host defense, hematopoiesis, granulomatous inflammation and type 1-type 2 cytokine balance in mice lacking CC chemokine receptor 1. J Exp Med 185:1959–1968

Gao W, Topham PS, King JA, Smiley ST, Csizmadia V, Lu B, Gerard CJ, Hancock WW (2000) Targeting of the chemokine receptor CCR1 suppresses development of acute and chronic cardiac allograft rejection. J Clin Invest 105:35–44

Garred P, Madsen HO, Petersen J, Marquart H, Hansen TM, Sorensen SF, Volck B, Svejgaard A, Andersen V (1998) CC chemokine receptor 5 polymorphism in rheumatoid arthritis. J Rheumatol 25:1462–1465

Gerard C, Frossard JL, Bhatia M, Saluja A, Gerard NP, Lu B, Steer M (1997) Targeted disruption of the b-chemokine receptor CCR1 protects against pancreatitis-associated lung injury. J Clin Invest 100:2022–2027

Gerard G, Rollins BJ (2001) Chemokines and disease. Nature Immunol 2:108–115

Hall IP, Wheatley A, Christie G, McDougall C, Hubbard R, Helms PJ (1999) Association of CCR5 delta32 with reduced risk of asthma. Lancet 354:1264–1265

Hancock WW, Lu B, Gao W, Csizmadia V, Faia K, King JA, Smiley ST, Ling M, Gerard NP, Gerard C (2000) Requirement of the chemokine receptor CXCR3 for acute allograft rejection. J Exp Med 192:1515–20

Heath H, Qin SX, Rao P, Wu LJ, LaRosa G, Kassam N, Ponath PD, Mackay CR (1997) Chemokine receptor usage by human eosinophils – The importance of CCR3 demonstrated using an antagonistic monoclonal antibody. J Clin Invest 99:178–184

Hébert CA, Vitangcol RV, Baker JB (1991) Scanning mutagenesis of interleukin-8 identifies a cluster of residues required for receptor binding. J Biol Chem 266:18989–18994

Hendrix CW, Flexner C, MacFarland RT, Giandomenico C, Fuchs EJ, Redpath E, Bridger G, Henson GW (2000) Pharmacokinetics and safety of AMD-3100 a novel antagonist of the CXCR-4 chemokine receptor, in human volunteers. Antimicrob Agents Chemother 44:1667–1673

Hesselgesser J, Ng HP, Liang M, Zheng W, May K, Bauman JG, Monahan S, Islam I, Wei GP, Ghannam A, Taub DD, Rosser M, Snider RM, Morrissey MM, Perez HD, Horuk R (1998) Identification and characterization of small molecule functional antagonists of the CCR1 chemokine receptor. J Biol Chem 273:15687–15692

Horuk R, Clayberger C, Krensky AM, Wang Z, Grone, H-J, Weber C, Weber KSC, Nelson PJ, May K, Rosser M, Dunning L, Liang M, Buckman B, Ghannam A, Ng HP, Islam I, Bauman JG, Wei, G-P, Monahan S, Xu W, Snider RM, Morrissey MM, Hesselgesser J, Perez HD (2001a) A nonpeptide functional antagonist of the CCR1 chemokine receptor is effective in rat heart transplant rejection. J Biol Chem 276:4199–4204

Horuk R, Ng HP (2000) Chemokine receptor antagonists. Med Res Rev 20:155–168

Horuk R, Shurey S, Ng HP, May K, Bauman JG, Islam I, Ghannam A, Buckman B, Wei GP, Xu W, Liang M, Rosser M, Dunning L, Hesselgesser J, Snider RM, Morrissey MM, Perez HD, Green C (2001b) CCR1-specific nonpeptide antagonist: efficacy in a rabbit allograft rejection model. Immunol Lett 76:193–201

Izikson L, Klein RS, Charo IF, Weiner HL, Luster AD (2000) Resistance to experimental autoimmune encephalomyelitis in mice lacking the CC chemokine receptor (CCR)2. J Exp Med 192:1075–1080

Lapierre JM, Morgans D, Wilhelm R, Mirzadegan TR (1998) The synthesis and biological evaluation of potent MCP-1 inhibitors. 26th National Medicinal Chemistry Symposium, Richmond, VA

Liang M, Mallari C, Rosser M, Ng HP, May K, Monahan S, Bauman JG, Islam I, Ghannam A, Buckman B, Shaw K, Wei GP, Xu W, Zhao Z, Ho E, Shen J, Oanh H, Subramanyam B, Vergona R, Taub D, Dunning L, Harvey S, Snider RM, Hesselgesser J, Morrissey MM, Perez HD, Horuk R

(2000) Identification and characterization of a potent, selective, and orally active antagonist of the CC chemokine receptor-1. J Biol Chem 275:19000–19008

Liu R, Paxton WA, Choe S, Ceradini D, Martin SR, Horuk R, MacDonald ME, Stuhlmann H, Koup RA, Landau NR (1996) Homozygous defect in HIV-1 Coreceptor accounts for resistance of some multiply-exposed individuals to HIV-1 infection. Cell 86:367–377

Mackay CR (2001) Chemokines:immunology's high impact factors. Nature Immunol 2:95–101

Mirzadegan T, Diehl F, Ebi B, Bhakta S, Polsky I, McCarley D, Mulkins M, Weatherhead GS, Lapierre JM, Dankwardt J, Morgans Jr D, Wilhelm R, Jarnagin K (2000) Identification of the binding site for a novel class of CCR2b chemokine receptor antagonists. binding to a common chemokine receptor motif within the helical bundle. J Biol Chem 275:25562–25571

Muller A, Homey B, Soto H, Ge N, Catron D, Buchanan ME, McClanahan T, Murphy E, Yuan W, Wagner SN, Barrera JL, Mohar A, Verastegui E, Zlotnik A (2001) Involvement of chemokine receptors in breast cancer metastasis. Nature 410:50–56

Murakami T, Nakajima T, Koyanagi Y, Tachibana K, Fujii N, Tamamura H, Yoshida N, Waki M, Matsumoto A, Yoshie O, Kishimoto T, Yamamoto N, Nagasawa T (1997) A small molecule CXCR4 inhibitor that blocks T cell line-tropic HIV-1 infection. J Exp Med 186:1389–1393

Murphy PM, Baggiolini M, Charo IF, Hebert CA, Horuk R, Matsushima K, Miller LH, Oppenheim JJ, Power CA (2000) International union of pharmacology. XXII. Nomenclature for chemokine receptors. Pharmacol Rev 52:145–176

Naya A, Kobayashi K, Ishikawa M, Ohwaki K, Saeki T, Noguchi K, Ohtake N (2001) Discovery of a novel CCR3 selective antagonist. Bioorg Med Chem Lett 11:1219–1223

Paxton WA, Martin SR, Tse D, O'Brien TR, Skurnick J, VanDevanter NL, Padian N, Braun JF, Kotler DP, Wolinsky SM, Koup RA (1996) Relative resistance to HIV infection of CD4 lymphocytes from persons who remain uninfected despite multiple high-risk sexual exposures. Nat Med 2:412–417

Reyes, G (2001) Development of CCR5 antagonists as a new class of anti-HIV therapeutic. 8th Conference on Retroviruses and Opportunistic Infections, Chicago, Illinois

Rothenberg ME, MacLean JA, Pearlman E, Luster AD, Leder P (1997) Targeted disruption of the chemokine eotaxin partially reduces antigen-induced tissue eosinophilia. J Exp Med 185:785–90

Rottman JB, Slavin AJ, Silva R, Weiner HL, Gerard CG, Hancock WW (2000) Leukocyte recruitment during onset of experimental allergic encephalomyelitis is CCR1 dependent. Eur J Immunol 30:2372–7

Samson M, Libert F, Doranz BJ, Rucker J, Liesnard C, Farber CM, Saragosti S, Lapouméroulie C, Cognaux J, Forceille C, Muyldermans G, Verhofstede C, Burtonboy G, Georges M, Imai T, Rana S, Yi YJ, Smyth RJ, Collman RG, Doms RW, Vassart G, Parmentier M (1996) Resistance to HIV-1 infection in Caucasian individuals bearing mutant alleles of the CCR-5 chemokine receptor gene. Nature 382:722–725

Schall TJ, Dairaghi DJ, McMAster BE (2001) Compounds and methods for modulating CXCR3 function. In World (PCT) Patent W0–161114

Schols D, Struyf S, Damme JV, Est JA, Henson G, Clercq ED (1997) Inhibition of T-tropic HIV strains by selective antagonization of the chemokine receptor CXCR4. J Exp Med 186:1383–1388

Sevenet T (1991) Looking for new drugs: what criteria? J Ethnopharmacol 32:83–90

Shiota T, Yamagami S, Kataoka K, Endo N, Tanaka H, Barnum D, Greene J, Moree W, Ramirez-Weinhouse M, Tarby, C (1997) Diarylalkyl cyclic diamine derivatives as chemokine receptor antagonists. In World (PCT) Patent WO-09744329

Strieter RM, Polverini PJ, Kunkel SL, Arenberg DA, Burdick MD, Kasper J, Dzuiba J, Van Damme J, Walz A, Marriott D, Chan SY, Roczniak S, Shanafelt AB (1995) The functional role of the ELR motif in CXC chemokine-mediated angiogenesis. J Biol Chem 270:27348–27357

Tachibana K, Hirota S, Iizasa H, Yoshida H, Kawabata K, Kataoka Y, Kitamura Y, Matsushima K, Yoshida N, Nishikawa A, Kishimoto T, Nagasawa T (1998) The chemokine receptor CXCR4 is essential for the vascularization of the gastrointestinal tract. Nature 393:591–594

Thelen M (2001) Dancing to the tune of chemokines. Nature Immunol 2:129–34

White JR, Lee JM, Young PR, Hertzberg RP, Jurewicz AJ, Chaikin MA, Widdowson K, Foley JJ, Martin LD, Griswold DE, Sarau HM (1998) Identification of a potent, selective nonpeptide CXCR2 antagonist that inhibits interleukin-8-induced neutrophil migration. J Biol Chem 273:10095–10098

Yang Y, Loy J, Ryseck RP, Carrasco D, Bravo R (1998) Antigen-induced eosinophilic lung inflammation develops in mice deficient in chemokine eotaxin. Blood 92:3912–23

Ying S, Meng Q, Zeibecoglou K, Robinson DS, Macfarlane A, Huer M, Kay AB (1999) Eosinophil chemotactic chemokines (Eotaxin, Eotaxin-2, RANTES, monocyte chemoattractant protein-3 (MCP-3), and MCP-4), and C-C chemokine receptor 3 expression in bronchial biopsies from atopic and nonatopic (intrinsic) asthmatics. J Immunol 163:6321–6329

11 The Role of Chemokines in Inflammatory Skin Diseases

G. Girolomoni, S. Pastore, A. Cavani, C. Albanesi

11.1 Introduction

The skin provides a major boundary between the host and the external environment, and it is obviously very well equipped to mount effective immune responses against microorganisms. Dendritic cells (DCs), including epidermal Langerhans cells and dermal DCs, specialize in recognizing and capturing foreign antigens as well as in the activation of naive T cells, and are thus essential for the induction of immune responses. T lymphocytes transduce antigen recognition into effector mechanisms to eliminate pathogens. The recruitment of T cells and other leukocytes at the site of skin inflammation is therefore a critical step for an efficient engagement of potentially dangerous signals (Koelle et al. 2002). The outcome of T cell-dependent skin reactions also depends on the cross-talk between infiltrating T cells and resident skin populations. Keratinocytes, mast cells, and endothelial cells, which constitute the static component of the skin immune system, as well as DCs contribute to attract discrete T cell subsets in the skin. In turn, activated T cells secrete cytokines and express molecules which activate resident cells, leading to amplification of the inflammatory reaction.

As an inevitable counterpart of its location and duty to prepare efficient immune responses, the skin is also a frequent site of hypersensitivity reactions against apparently harmless antigens, such as the haptens causing allergic contact dermatitis (ACD). Moreover, the skin is subjected to complex immunopathological reactions having a strong genetic component. The most common of these genetically-determined inflammatory disorders are atopic dermatitis (AD) and psoriasis. Valuable animal models of ACD are available and the disease can be easily induced in humans, allowing a detailed analysis of the cell types and molecules involved in the induction, expression, and regulation of this immune reaction. Thus, ACD has represented a prototype for T cell-mediated skin immune responses, and many of the findings observed in this reaction have been extended to other disorders. In contrast, no satisfying animal models have been developed for atopic dermatitis and psoriasis, rendering difficult the study of the various components involved in the initiation and expression of these diseases.

11.2 Chemokines and Chemokine Receptors

Chemokines are a superfamily of structurally related, small (6–14 kDa) proteins that regulate the traffic of various types of leukocytes, including lymphocytes, DCs, monocytes, neutrophils and eosinophils (Sallusto et al. 2000; Zlotnik and Yoshie 2000; Mackay 2001; Moser and Loetscher 2001; Kunkel and Butcher 2002; Luster 2002). Advances in molecular cloning techniques as well as the development of expressed sequence tag databases allowed the rapid identification of a vast array of chemokine-related genes, and so far more than 50 human chemokines have been characterized (http://cytokine.medic.kumamoto-u.ac.jp/). Chemokines are classified in four subfamilies, according to the position of two highly conserved

Table 1. Principal human chemokines and chemokine receptors

Systematic name	Current name	Receptor(s)
CXC family		
CXCL1	Growth-related oncogene (GRO)α, MGSA-α, NAP-3	CXCR2
CXCL2	GROβ, MGSA-β, MIP-2α	CXCR2
CXCL3	GROγ, MGSA-γ, MIP-2β	CXCR2
CXCL4	PF4	Unknown
CXCL5	ENA-78	CXCR1, CXCR2
CXCL6	GCP-2, CKA-3	CXCR1
CXCL7	NAP-2	CXCR2
CXCL8	IL-8	CXCR1, CXCR2
CXCL9	Monokine-induced by interferon γ (Mig)	CXCR3, *CCR3**
CXCL10	Interferon-induced protein of 10 kDa (IP-10), C7	CXCR3, *CCR3**
CXCL11	Interferon-induced T cell α-chemo-attractant (I-TAC), IP-9	CXCR3, *CCR3**
CXCL12	Stromal-derived factor (SDF)-1α/β	CXCR4
CXCL13	B cell-activating chemokine (BCA)-1, BLC	CXCR5
CXCL14	Breast and kidney chemokine (BRAK), bolekine, MIP-2γ	Unknown
CXCL15	Lungkine	Unknown
CXCL16	SR-PSOX	CXCR6
CC family		
CCL1	I-309	CCR8
CCL2	Monocyte chemotactic protein (MCP)-1, MCAF, SMC-CF	CCR1, 2
CCL3L1	LD78β	CCR1, CCR5
CCL3	Macrophage inflammatory protein (MIP)-1α, LD78α, SISα	CCR1, CCR5
CCL4	MIP-1β, LAG-1	CCR1, CCR5
CCL5	Regulation and activated normal T cell expressed and secreted (RANTES)	CCR1, CCR3, CCR5
CCL7	MCP-3	CCR1, 2, 3, *CCR5**
CCL8	MCP-2	CCR1, 2, 3, 4, 5
CCL11	Eotaxin-1	CCR3, CCR5, *CXCR3**, *CCR2**
CCL13	MCP-4, NCC-1	CCR1, 2, 3, CCR5

Table 1 (continued)

Systematic name	Current name	Receptor(s)
CCL14	HCC-1, HCC-3, NCC-2	CCR1, 3, 5
CCL15	HCC-2, Lkn-1, MIP-1δ, NCC-3	CCR1, 3
CCL16	HCC-4, LEC, NCC-4	CCR1
CCL17	Thymus- and activation-regulated chemokine (TARC)	CCR4
CCL18	Dendritic cell chemokine 1 (DC-CK1), PARC, AMAC-1, MIP-4	Unknown
CCL19	MIP-3β, ELC, exodus-3	CCR7, 11
CCL20	MIP-3α, LARC, exodus-1, ST-38	CCR6
CCL21	Secondary lymphoid tissue chemokine (SLC), 6Ckine, exodus-2	CXCR3, CCR7, CCR11
CCL22	Macrophage-derived chemokine (MDC), STCP-1	CCR4
CCL23	MPIF-1	CCR1
CCL24	Eotaxin-2, MPIF-2,	CCR3
CCL25	Thymus-expressed chemokine (TECK)	CCR9, 11
CCL26	Eotaxin-3, MIP-4α	CCR3
CCL27	Cutaneous T cell-attracting chemokine (CTACK), ILC	CCR10
CCL28	Mucosae-associated epithelial chemokine (MEC)	CCR10
C family		
XCL1	Lymphotactin, SCM-1α, ATAC	XCR1
XCL2	SCM-1β	XCR1
CXXXC family		
CX3CL1	Fractalkine	CX$_3$CR1

Chemokine binding to the receptors in italics (*) antagonizes natural agonistic chemokines.

cysteine residues at the N-terminus of the molecule (Table 1). Upon secretion, chemokines bind to extracellular matrix and cell membrane proteoglycans forming stable gradients near to the site of production. CX3CL1/Fractalkine is an exception as it is expressed as a membrane integral protein; however, it can be shed from the membrane (Bazan et al. 1997). Although their main function is to regulate cell trafficking, chemokines also display important roles in governing leukocyte activation and differentiation (Sallusto et al. 1999a;

Luther and Cyster 2001). The specific effects of chemokines on target cells are mediated by seven transmembrane spanning, G-protein coupled receptors (Rossi and Zlotnik 2000; Mellado et al. 2001). To date, about 20 chemokine receptor-encoding genes have been identified, mostly located in clusters on chromosomes 2 and 3 (Table 1). The chemokine-chemokine receptor axis is highly promiscuous, allowing a single chemokine to bind different receptors and a chemokine receptor to transduce the signal for several chemokines. In contrast to this notion, some recently identified chemokines show a very high receptor and tissue specificity, and are thought to contribute to tissue-restricted leukocyte trafficking (Campbell et al. 1999; Zabel et al. 1999). In particular, CCL27/CTACK is predominantly expressed in the skin and bind specifically CCR10 (Morales et al. 1999; Homey et al. 2000b). Also the CCL17/TARC-CCR4 system has been proposed to be determinant for the homing of CCR4-bearing T lymphocytes to inflamed skin (Campbell et al. 1999). Moreover, the same chemokine may transduce different signals according to the bound receptor: for instance, CXCL9/Mig, CXCL10/IP-10, and CXCL11/I-TAC are agonists to the CXCR3 receptor but they also bind to the CCR3 receptor, antagonizing other CCR3 ligands (Loetscher et al. 2001). Similarly, CCL11/eotaxin is a natural agonist for CCR3 and CCR5 and an antagonist for CCR2 and CXCR3 (Uguccioni et al. 1997; Weng et al. 1998; Ogilvie et al. 2001). Depending on their function, regulation, and site of production, chemokines can be classified as "inflammatory" and "lymphoid" or "homeostatic." Inflammatory chemokines are expressed in inflamed tissues by resident or infiltrating cells, and recruit effector cells. Homeostatic chemokines, such as CCL19/ELC/MIP-3β, CCL21/SLC, and CXCL13/BCA1, are primarily produced within lymphoid tissues and are involved in the maintenance of the constitutive lymphocyte traffic and cell compartmentalization within these organs (Sallusto et al. 2000; Moser and Loetscher 2001; Luster 2002).

11.3 Chemokine Receptors on T Cell Subsets

T lymphocyte circulation in peripheral tissues encompasses a series of complex events including adhesion to endothelial cells mediated by selectins, integrins, and adhesion molecules and by the local expression of chemotactic stimuli (von Andrian and Mackay 2000). Chemokines are physiological activators of rapid lymphocyte arrest along high endothelial venules (HEV) in secondary lymphoid organs, and play a central role in lymphocyte tissue selective homing. Transient tethering and rolling precede firm adhesion of circulating leukocytes and are essential to slow leukocyte motion, thus facilitating microenvironmental sampling and subsequent interaction with proadhesive chemokines presented by the endothelial cells. Tethering and rolling of leukocytes on vessel walls are primarily mediated by specialized selectins and mucins. These relatively loose adhesive interactions are rapidly converted into integrin-dependent firm adhesion upon chemokine receptor engagement and generation of intracellular signals. The functions of adhesion molecules and chemokine receptors are reciprocally regulated. Quantitative variations in chemokine receptor expression level and ligand engagement may alter the selectivity of integrin-dependent lymphocyte adhesive responses, suggesting a mechanism by which chemokine networks may either generate or break the specificity of lymphocyte subset recruitment (D'Ambrosio et al. 2002). On the other hand, adhesion molecule triggering changes chemokine receptor expression.

Chemokine receptor profile on T lymphocytes varies depending on their activation stage, differentiation pattern, and tissue targeting. Naive T lymphocytes mostly home to lymph nodes, where they are experienced by antigen-loaded DCs migrated from peripheral tissues. Lymph node entry of naive T cells is regulated by the expression of CCR7, which binds SLC produced by the endothelial cells of the HEV (Gunn et al. 1998). CCR7-SLC interaction increases $\alpha_L\beta_2$ integrin adhesiveness for ICAM-1/2 on endothelial cells, thus promoting T lymphocyte firm adhesion and transmigration. Once T cells have crossed the HEV, their interaction with DCs is favored by MIP-3β and CCL18/DC-CK1/PARC, which are secreted by DCs and attract naive T lymphocytes expressing the cognate receptor CCR7. In addition, DCs are the most abundant source of CCL22/MDC and TARC that bind the CCR4 receptor expressed by recently activated T cells (Sallusto et al. 2000; Vulcano et al. 2001). Once experienced, T lymphocytes completely rearrange their chemokine receptor profile and acquire new migratory capacity to allow their homing

Table 2. Chemokine receptors expressed by resting/activated CD4$^+$ T cell subsets

	CD4$^+$ T cells		
	Th1	Th2	Tr1
CCR1	++/+	++/+	++/+
CCR2	++/+	++/+	++/+
CCR3	–/–	+++/+	+/+
CCR4	±/+	+++/+++	++/++
CCR5	+++/+	+/±	++/+
CCR8	–/–	++/+	+++/++
CXCR1	+/–	+/–	+/–
CXCR2	–/–	–/–	–/–
CXCR3	+++/+	±/–	++/+
CXCR4	+/+	+/+	++/+
CX$_3$CR1	+++/nd	±/nd	nd

nd, not determined.

in peripheral tissues. Importantly, acquisition of discrete chemokine receptors parallels the differentiation and cytokine polarization of T cells (Table 2). T helper (Th) 1 lymphocytes are rich in CXCR3 and CCR5 receptors, whereas Th2 cells express CCR3, CCR4, and CCR8 (Bonecchi et al. 1998; D'Ambrosio et al. 1998; Sallusto et al. 2000). The recently identified T regulatory (Tr) cells, producing high amounts of interleukin (IL)-10, coexpress both Th1- and Th2-associated receptors, with high levels of CCR8 and moderate amounts of CCR7 (Cavani et al. 2001; Sebastiani et al. 2001). CCR8, together with CCR4, is also highly expressed on the recently identified CD4$^+$CD25$^+$ regulatory T cells (Iellem et al. 2001). These findings suggest that the CCR8/I-309 axis may have a central role in the recruitment of multiple T cell populations involved in the control of peripheral immune responses. In contrast to the large body of information regarding the migratory behavior of CD4$^+$ T cells, chemokine responsiveness of CD8$^+$ T lymphocytes has been less investigated. Recent data indicate that skin-homing Tc1 and Tc2 CD8$^+$ cells both express high amount of CXCR3, whose expression in CD4$^+$ cells is mainly limited to the Th1 subset, and promptly respond to IP-10 in chemotaxis assays. Conversely, Tc1 and Tc2 cells show lower expression of CCR4, when compared to Th1 and Th2 lymphocytes (Sebastiani et al. 2002). Thus, the local expression of chemokines during an inflammatory process may contribute to the selective accumulation of different T cell subsets. T cell activation strongly affects chemokine receptor expression, with most receptors for inflammatory chemokines downregulated, except for CCR4 and CCR8, which are transiently upregulated (D'Ambrosio et al. 1998; Sallusto et al. 1999b). Receptor downregulation promotes a switch from a migratory to a stationary behavior so that T lymphocytes can better exert their effector activities. Besides naive T cells, a pool of central memory T cells express CCR7 (Sallusto et al. 1999c). Once activated, these lymphocytes upregulate CCR7 and migrate to lymph nodes where they can provide strong help for DC maturation and differentiate into effector cells.

The existence of subsets of memory T cells that preferentially migrate to the skin is well documented. Most of the skin-homing T lymphocytes express the cutaneous lymphocyte-associated antigen (CLA), the ligand for E-selectin expressed by activated endothelial cells of the skin microvasculature (Robert and Kupper 1999). It has been recently described that skin-seeking CD4$^+$ T lymphocytes express CCR4 independently from their cytokine releasing profile. CCR4 permits T lymphocyte adhesion to TARC, which is exposed on the surface of activated endothelium at the site of skin inflammation (Campbell et al. 1999). CCR4 triggering by TARC induces T lymphocyte integrin activation and promotes the firm adhesion of circulating lymphocytes to the endothelium. CCR6, albeit also expressed in gut migrating lymphocytes, has been described as an important chemokine receptor that favors T cell migration into the skin (Liao et al. 1999). In particular, tumor necrosis factor (TNF)-α-activated dermal microvascular endothelial cells produce large amounts of CCL20/MIP-3α which is critical for arrest of CCR6$^+$ memory T cells (Fitzhugh et al. 2000). Finally, CTACK is a newly identified tissue-specific chemokine which is predominantly produced by epidermal keratinocytes, and recruits CCR10$^+$ skin-homing T lymphocytes in inflamed skin (Morales et al. 1999; Homey et al. 2000b, 2002).

11.4 Chemokines Production by Resident Skin Cells

Resident skin cells greatly contribute to T lymphocyte recruitment during inflammation. A limited amount of some chemokines is expressed in unperturbed skin, and is probably involved in the basal trafficking of memory T cells and DCs, although no conclusive data are available. Keratinocytes are a relevant source of chemokines, with resting cells releasing low levels of CXCL8/IL-8 and expressing CTACK and CXCL12/SDF-1 mRNA (Anttila et al. 1992; Morales et al. 1999). Constitutive keratinocyte production of MIP-3α has been indicated to be relevant to the basal recruitment of Langerhans cells in the epidermis (Charbonnier et al. 1999), although other authors have failed to show significant production by unstimulated keratinocytes and expression in normal epidermis (Dieu-Nosejan et al. 2000; Homey et al. 2000b; Nakayama et al. 2000). Following activation, keratinocytes rapidly upregulate IL-8 and CTACK, and synthesize CXCL1/Gro-α, IP-10, Mig, I-TAC, CCL5/RANTES, CCL2/MCP-1, MIP-3α, MDC, and CCL1/I-309. TARC is constitutively expressed at very low levels by primary keratinocytes, and its induction in response to cytokines has been shown by Vestergaard et al. (2000), but we were not able to confirm this finding (Albanesi et al. 2001; D'Ambrosio et al. 2002). IP-10, MCP-1, and IL-8 are the most abundant chemokines released by activated keratinocytes. In vitro investigations have indicated that inflammatory and T cell-derived cytokines are the most potent stimuli for keratinocyte activation. In particular, IL-1 and TNF-α strongly induce MIP-3α, RANTES, and IL-8 (Albanesi et al. 1999; Dieu-Nosejan et al. 2000; Nakayama et al. 2000). The type 1 cytokine interferon (IFN)-γ, alone or together with TNF-α, rapidly promotes the secretion of very high levels of CXCR3 agonists and MCP-1 and low amounts of I-309 and MDC (Albanesi et al. 1999; Albanesi et al. 2001). IL-17, a cytokine produced by part of Th1 and Th2 lymphocytes, modulates the effects of IFN-γ on keratinocytes by augmenting IL-8, and decreasing RANTES release (Albanesi et al. 1999; Albanesi et al. 2000a). Notably, IL-4, a type 2 cytokine, enhances IFN-γ-induced release of IP-10, Mig, and I-TAC by cultured keratinocytes (Albanesi et al. 2000b), indicating that the Th2 to Th1 switch observed in some chronic inflammatory skin diseases (e.g., AD) could partly de-

pend on the predilection of keratinocytes to release Th1-active chemokines (Albanesi et al. 2001). Keratinocytes from genetically determined inflammatory skin disorders may have intrinsic abnormalities
also in their capacity to produce chemokines. For instance, keratinocytes cultured from AD patients show an exaggerated production of
RANTES, whereas psoriatic keratinocytes overproduce IL-8, MCP-1,
and IP-10 (Pastore et al. 2000; Giustizieri et al. 2001). These defects
may help to explain the accumulation of distinct leukocyte types in
these two diseases. Recent findings of our group have demonstrated
that the expression of the chemokines IP-10, Mig, and MCP-1 induced by IFN-γ in human keratinocytes can be quite abrogated by
overexpressing the negative regulators of IFN-γ signaling SOCS1
and SOCS3 (Federici et al. 2002). These latter proteins belong to a
family of intracellular proteins which regulate the magnitude and
duration of responses triggered by various cytokines by inhibiting
their signal transduction pathway in a classic negative feedback loop
(Yasukawa et al. 2000). SOCS molecules were strongly expressed by
keratinocytes in lesional skin of psoriasis and ACD patients, and to
a lesser extent in AD lesions. The inhibition of the IFN-γ-induced
IP-10, Mig, and MCP-1 expression observed in keratinocytes permanently transduced with SOCS1 and SOCS3 was dependent on the reduction of the IFN-γRα phosphorylation and, consequently, of
STAT1 and STAT3 protein activities. In contrast, IL-8 production by
keratinocyte clones overexpressing SOCS1 was enhanced and associated to an abrogation of IFN-γ-induced growth inhibition of these
cells. These findings identify SOCS1 and SOCS3 molecules as new
potential molecular targets for those IFN-γ-dependent skin diseases
where chemokines are aberrantly expressed by keratinocytes.

Other resident cells can contribute to chemokine production during skin inflammation. Dermal fibroblasts produce IL-8, RANTES,
MIP-3α, and Gro-α when stimulated with IL-1 and TNF-α (Schröder
and Mochizuki 1999; Dieu-Nosejan et al. 2000). Moreover, these
cells constitutively express the mRNA for CCL11/eotaxin, CCL24/
eotaxin-2 and CCL26/eotaxin-3, which are upregulated in response
to TNF-α and IL-4 (Mochizuchi et al. 1998; Dulkys et al. 2001).
Dermal fibroblasts also express CCL13/MCP-4 mRNA in response
to proinflammatory cytokines (Petering et al. 1998). All these chemokines are strong attractants for eosinophils and Th2 cells and may

play a more important role in AD and other diseases with eosinophil involvement (Schröder and Mochizuchi 1999). In addition to being major effector cells in the elicitation of allergic inflammation, mast cells may affect T-cell recruitment to sites of inflammation by releasing chemotactic stimuli such as IL-8, XCL1/lymphotactin and IL-16 (Mekory and Metcalfe 1999). DCs represent another relevant source of chemokines, whose production follows a precise sequential order during maturation (Sallusto et al. 1999c). After an initial burst of inflammatory chemokines (MIP-1α, MIP-1β, RANTES, MCP-1, and IL-8), maturing DCs produce lymphoid chemokines such as TARC, MDC, PARC, and ELC, the two latter active on naive T cells. Interestingly, different DC subsets appear to express a similar pattern of chemokines attracting T cells (Vissers et al. 2001). Extracellular nucleotides (e.g., ATP) may represent constitutive signals that can alert the immune system of abnormal cell death. Relatively high doses of nucleotides induce rapid release of proinflammatory mediators and favor pathogen killing. However, recent findings on antigen presenting cells, particularly DCs, revealed a more complex role for these molecules. Chronic exposure to low dose nucleotides can redirect cellular responses to prototypic activation stimuli, leading to suppressed inflammation and immune deviation. ATP affects the pattern of chemokine release from DCs by upregulating the constitutive production of MDC and inhibiting the lipopolysaccharide (LPS)-induced secretion of IP-10 and RANTES. This results in a selectively impaired capacity of DCs to recruit type 1 but not type 2 lymphocytes (la Sala et al. 2002). In contrast, IFN-γ-treated DCs markedly upregulate the release of IP-10 and significantly reduce the secretion of TARC, attracting preferentially Th1 and Th2 cells, respectively. Together with an upregulation of IL-12 release, IFN-γ-induced increase of IP-10 may thus set DCs for a more effective Th1 polarization of the immune response (Corinti et al. 2002). DCs are the largest source of MDC, and in situ MDC staining is most evident in mature CD83$^+$ DCs, with the number of MDC$^+$ DCs much higher in AD lesional skin than in the skin affected with psoriasis or ACD (Vulcano et al. 2001). In the complex process of leukocyte recruitment, some chemokines have been shown to induce firm arrest via integrins both in vitro and in vivo. Upon exposure to inflammatory signals, endothelial cells express MIP-3α, SDF-1,

SLC, TARC, MCP-1, IL-8, RANTES, Gro-α, and MIP-1β, some of which are involved in the arrest of lymphocytes under physiologic flow conditions (von Andrian and Mackay 2000). In particular, TARC has been detected in venules of inflamed human skin and determines adhesion of skin memory T cells to ICAM-1 in vitro (Campbell et al. 1999). Moreover, Fitzhugh et al. (2000) demonstrated that MIP-3α produced by endothelial cells may be critical for the arrest of CCR6$^+$ T lymphocytes on activated dermal endothelium, suggesting a role for this chemokine in T cell recruitment in psoriatic skin where CCR6 is upregulated. Finally, fractalkine is expressed on the membrane of endothelial cells and modulated by inflammatory signals and cytokines. The fractalkine/CX3CR1 axis can be part of an amplification circuit of polarized type 1 responses, as suggested by the higher expression of this chemokine in psoriasis than in AD (Fraticelli et al. 2001).

11.5 Chemokines and Allergic Contact Dermatitis

ACD is a T cell-mediated inflammatory reaction occurring in sensitized individuals at the site of hapten challenge. In the sensitization phase, hapten penetrated into the skin is picked up by skin DCs which then migrate to the draining lymph nodes and present the hapten-peptide-MHC complexes to naive antigen-specific T lymphocytes. This process leads to the clonal expansion of specific T cells which can then be recruited to the skin. The migration of skin DCs to the lymph nodes is regulated by various factors, including a profound rearrangement in the pattern of chemokine receptors, which allow DCs to encounter naive T cells (Sozzani et al. 2000; Rennert et al. 2001). In mice deficient in CCR7 or its ligand, SLC, skin DCs do not migrate to the lymph nodes and contact hypersensitivity (CH) does not develop (Förster et al. 1999; Gunn et al. 1999). In sensitized subjects, a re-exposure to the antigen determines the activation and expansion of specific memory T cells. The inflammatory infiltrate in ACD is mainly composed of CD4$^+$and CD8$^+$ T cells, monocytes, and DCs, with an early and transient presence of neutrophils. The expression of both murine CH and human ACD correlates with the activity of hapten-specific CD8$^+$ T cells, which exert their effector function through direct cytotoxic activity as well as the release of cytokines (Bouloc et al, 1998; Kehren et al. 1999; Traidl et al. 2000; Trautman et al. 2000; Wang et al. 2000). On the other hand, $CD4^+$ T cells play a more complex role, with Th1 and Th2 subsets contributing to disease expression, and Tr cells primarily involved in its modulation and/or termination (Cavani et al. 2001; Girolomoni et al. 2001). In addition, Th2 cells may also exert a regulatory role as IL-4 administration or passive transfer of hapten-specific Th2 lymphocytes can reverse established murine CH (Biedermann et al. 2001). However, CH is reduced or unmodified in IL-4$^{-/-}$ mice and decreased in STAT-6 deficient mice in which Th2 responses are impaired (Yokozeki et al. 2000).

During ACD, leukocyte recruitment is under the control of the sequential and coordinated release of chemokines from resident and immigrating cells. In a recent work, Goebeler et al. (2001) analyzed the expression pattern of chemokines in the skin at different time points after hapten application. By using in situ hybridization, they showed that MCP-1 is already detectable at 6 h after hapten challenge, whereas RANTES and MDC appear at 12 h concomitantly with the infiltration of mononuclear cells in the dermis and epidermis. The expression of IP-10, Mig, TARC, and PARC began at 12 h and peaked at 72 h, paralleling the strong infiltration of lymphocytes. The variety of chemokines expressed during ACD determines a robust and rapid recruitment of leukocytes into the skin. MCP-1 seems to play a relevant role in ACD. In fact, transgenic mice overexpressing MCP-1 in basal keratinocytes showed enhanced CH responses together with an increased number of infiltrating DCs (Nakamura et al. 1995). It is still unclear which chemokine(s) drives the initial influx of T cells, the cellular source(s), and the nature of the induction stimulus. Keratinocytes may represent important contributors to this phase, although a TNF-α-mediated induction by haptens has been demonstrated only for IL-8 (Griffiths et al. 1991). Activated endothelial cells can conceivably contribute to early leukocyte arrival in the skin, by expressing both adhesion molecules and chemokines (Campbell et al. 1999; Fitzhugh et al. 2000; von Andrian and Mackay 2000). During the elicitation of ACD, upregulated expression of epidermal CTACK is observed already after 6 h. After 24 h, CTACK immunoreactivity is also detectable free in the papillary dermis and on dermal microvessels, most likely as the consequences of secretion from keratinocytes followed by absorption on extracellular matrix proteins and endothelial cells, respectively. At 24–48 h, the number of CCR10[+] T cells increases dramatically first in the perivascular and subepidermal areas and then in the epidermis (Homey et al. 2002). Also mast cells and platelets might be indirectly involved in this early T cell recruitment through the release of serotonin, which together with TNF-α activate the endothelium facilitating cell entrance. This phenomenon seems to be C5a-dependent, since C5a knockout mice do not show T cell infiltration at the sites of hapten challenge (Tsuji et al. 2000). Moreover, mast cells can be importantly involved in neutrophil recruitment through the release of TNF-α and MIP-2, the functional analogue of human IL-8, in murine CH (Biedermann et al. 2000). Infiltrating monocytes and DCs are themselves a source of chemokines for successive boosts of lymphocyte arrival. The activation of some hapten-specific T lymphocytes into the skin leads to the production of cytokines like IFN-γ, TNF-α, and IL-4 which in turn stimulate keratinocytes and other cells to produce IP-10, Mig, and I-TAC, the ligands for CXCR3 (Albanesi et al. 1999; Albanesi et al. 2000b; Albanesi et al. 2001). These chemokines selectively attract T lymphocytes, which then rapidly accumulate in the epidermis. Indeed, keratinocytes, continuously stimulated by T cell-derived cytokines, produce large amounts of CXCR3 ligands thus contributing to further accumulation of CXCR3-bearing T cells. The result is that more than 70% of ACD infiltrating T lymphocytes are CXCR3[+] (Flier et al. 1999; Albanesi et al. 2000b). The higher expression of CXCR3 and of CCR4 on skin-homing nickel-specific CD8[+] and CD4[+] T cells suggests that trafficking of these two populations at the site of skin inflammation differs, with CD4[+] and CD8[+] cells recruitment primarily directed by the CCR4/TARC and CXCR3/IP-10 axis, respectively (Sebastiani et al. 2002). Once recruited into the skin, activated T lymphocytes represent a relevant source of chemokines. Upon T cell receptor triggering, Th1 and Th2 cells produce similar sets of chemokines, including RANTES, I-309, IL-8, and MDC (Sallusto et al. 1999a). Some differences seem to exist in terms of chemokines production between CD4[+] and CD8[+] T lymphocytes. In a mouse model, it has been shown that IP-10 expression is primarily mediated by CD8[+]

and inhibited by CD4$^+$ T lymphocytes during the elicitation phase of CH (Abe et al. 1996).

Resolution of ACD is likely due to multiple mechanisms, including induction of T cell anergy and active suppression, and may involve several cell types. In this process, IL-10-producing T cells (Tr lymphocytes) might play a central role (Cavani et al. 2000; Schwarz et al. 2000; Cavani et al. 2001). We recently demonstrated that Tr cells migrate in vitro in response to various chemokines including MCP-1, MIPs, and TARC. More interestingly, I-309, which is not active on Th1 cells, attracts Tr lymphocytes more vigorously than Th2 cells. Consistent with these results, Tr lymphocytes express higher levels of CCR8 compared to Th2 cells. I-309 is produced by both keratinocytes and activated T cells, and is expressed in ACD skin with an earlier kinetics compared to IL-4 and IL-10 (Sebastiani et al. 2001). These data indicate that I-309/CCR8 may contribute relevantly to the termination of ACD through the recruitment of Tr lymphocytes. Also CCR6, the receptor of MIP-3a, seems to be involved in the modulation of CH as CCR6 deficient mice show increased and persistent CH responses, probably in relation to an impaired recruitment of CD4$^+$ regulatory cells (Varona et al. 2001).

11.6 Chemokines and Psoriasis

Psoriasis is a genetically determined skin disease characterized by aberrant proliferation and differentiation of keratinocytes as well as cutaneous inflammation. T cell-mediated immune mechanisms have a primary role in the pathogenesis of psoriasis (Nickoloff et al. 2000; Asadullah et al. 2002; Krueger 2002). In particular, activated Th1 cells releasing IFN-γ and TNF-a stimulate keratinocytes to produce cytokines, chemokines, and adhesion molecules, which further amplify the inflammatory response. Various studies have documented a strong chemokine expression in psoriatic keratinocytes in lesional skin, and keratinocyte production of chemokines may contribute relevantly to the establishment of the inflammatory infiltrate (Schön and Ruzicka 2001). Specifically, IL-8 and related chemokines are responsible for the intraepidermal collection of neutrophils (Anttila et al. 1992; Nickoloff et al. 1994). MCP-1, RANTES, IP-10, and other CXCR3 ligands attract predominantly monocytes and Th1 cells (Gillitzer et al. 1993; Gottlieb et al. 1998), whereas MIP-3α recruits immature Langerhans cells and DCs, and CLA$^+$ T cells (Dieu-Nosjean et al. 2000; Homey et al. 2000a). In line with these observations, T cells bearing CCR4, CXCR3, and CCR6 receptors are well represented in psoriatic skin lesions (Homey et al. 2000a; Rottman et al. 2001). In particular, CXCR3$^+$CD8$^+$ T cells are increased by tenfold in psoriatic epidermis compared with the frequency of these cells in peripheral blood of patients with psoriasis (Rottman et al. 2001). MCP-4 is also strongly expressed in the basal layers of the psoriatic epidermis and together with MIP-3α can direct the traffic of immature DCs (Vanbervliet et al. 2002). CTACK is abundantly present in basal and suprabasal keratinocytes of psoriatic lesions as well as in the dermis, together with a high number of CCR10$^+$ T cells (Homey et al. 2002). Consistent with the Th1-dominated immunity underlying psoriasis, CCR10 is preferentially expressed by skin-homing CLA$^+$ memory T cells secreting TNF and IFN-γ, but minimal IL-10 and IL-4 upon activation (Hudak et al. 2002). Activation of endothelial cells may represent an early event in the pathogenesis of psoriatic lesions, and allow the initial accumulation of T cells, which in turn can amplify the inflammatory response by releasing cytokines. Expression of MIP-3α on dermal endothelial cells may thus have an important role in the arrest of CCR6$^+$ immature DCs and memory T cells (Fitzhugh et al. 2000; Homey et al. 2000a), whereas expression of TARC is important for the arrest of CCR4$^+$ T cells (D'Ambrosio et al. 2002). Dermal endothelial cells of psoriasis lesions, but not those of AD, are strongly positive for fractalkine, a membrane-bound chemokine which is induced by IFN-γ and whose receptor (CXCR1) is preferentially expressed by Th1 cells and NK cells (Fraticelli et al. 2001).

The genetic predisposition to psoriasis may include an altered control of inflammatory gene expression in the skin. In particular, psoriatic keratinocytes may have intrinsic defects leading to exaggerated synthesis of certain chemokines such as IL-8, MCP-1, and IP-10 (Nickoloff et al. 1994; Giustizieri et al. 2001; Giustizieri et al. 2002), and also display increased expression of IL-8R, which can mediate an increased proliferative response of keratinocytes to the chemokine (Kulke et al. 1998; Schön and Ruzicka 2001). Moreover, psoriatic keratinocytes activated in vitro with IFN-γ and TNF-α showed an ICAM-1 induction higher than normal keratinocytes. The signal transduction initiated by IFN-γ and TNF-α involves principally a cooperation between STAT-1 and NF-κB transcription factors. In contrast, AP-1 is known to be less important in the signaling elicited by these cytokines. We have shown that in transiently transfected keratinocytes, IFN-γ and TNF-α induced a strong NF-κB and STAT-1 binding activity, whereas the induction of AP-1 function was less evident. Interestingly enough, psoriatic keratinocytes exhibited a more prominent NF-κB and STAT-1, but not AP-1 activity compared to control keratinocytes (Giustizieri et al. 2002). Indeed, perturbation in signal transduction pathways and in the activation of transcription factors has been implicated in the dysregulated functions of psoriatic keratinocytes (Karvonen et al. 2000; Haase et al. 2001).

Increasing evidence indicates that nitric oxide (NO) is involved in the maintenance of skin homeostasis as well as in the modulation of inflammatory reactions. NO is a short-lived radical produced from the L-arginine pathway by different isoforms of NO synthase (NOS), which are expressed by various cell types residing in the skin. High levels of NO have been measured in the skin affected with psoriasis, AD, or ACD (Sirsjo et al. 1996; Ormerod et al. 1998). In these conditions, proinflammatory cytokines stimulate keratinocytes to express inducible NOS (iNOS), which in turn catalyzes NO production. Fibroblasts and DCs also become iNOS positive after exposure to bacterial endotoxin and IFN-γ, and endothelial cells express iNOS after activation with IL-1β. The role of NO in the regulation of inflammatory responses has been extensively investigated. Depending on the concentration, the cell type and its state of activation, as well as the presence of other inflammatory mediators, NO can either block or stimulate inflammatory responses (Bodgan 2001). A novel function of NO is its ability to modulate chemokine expression. Recently, we tested whether synthetic NO donors could modulate the expression of chemokines and ICAM-1 in keratinocyte primary cultures established from healthy subjects and patients with psoriasis. NO donors (S-nitrosoglutathione and NOR-1) diminished in a dose-dependent manner and at both mRNA and protein levels IP-10, RANTES, and MCP-1 expression in keratinocytes cultured from healthy subjects and psoriatic patients. In contrast, constitutive and induced IL-8 production was unchanged. NO-treated psoriatic skin showed reduction of IP-10, RANTES, and MCP-1, but not IL-8 expression by keratinocytes. Moreover, the number of $CD14^+$ and $CD3^+$ cells infiltrating the epidermis and papillary dermis diminished significantly. NO donors also downregulated ICAM-1 protein expression without affecting mRNA accumulation in vitro, and suppressed keratinocyte ICAM-1 in vivo. Finally, NO donors inhibited NF-κB and STAT-1, but not AP-1 activities in transiently transfected keratinocytes. These results define NO donors as negative regulators of chemokine production by keratinocytes both in vitro and in vivo (Giustizieri et al. 2002). The clinical efficacy of novel targeted immunomodulatory therapies of psoriasis, such as IL-10 and dimethylfumarate, is associated with a downregulation of chemokine production and signaling pathway. In particular, administration of IL-10 to patients with chronic plaque psoriasis inhibited the epidermal IL-8 pathway by reducing the expression of IL-8, its receptor CXCR2, and its inducer IL-17 (Asadullah et al. 2001; Reich et al. 2001 a). Dimethylfumarate was shown to suppress the IFN-γ-induced production of Gro-α, IL-8, Mig, IP-10, and I-TAC in keratinocytes and peripheral blood mononuclear cells (Stoof et al. 2001).

11.7 Chemokines and Atopic Dermatitis

AD is a chronic inflammatory disease which results from complex interactions between genetic and environmental factors (Leung and Soter 2001). An altered lipid composition of the stratum corneum is responsible for the xerotic aspect of the skin, and may determine a higher permeability to allergens and irritants. Specific immune responses against a variety of environmental allergens are also implicated in AD pathogenesis with a bias towards Th2 immune responses. In particular, DCs expressing membrane IgE receptors play a critical role in the amplification of allergen-specific T cell responses (von Bubnoff et al. 2001; Kraft et al. 2002). Cross-linkage of specific IgE receptors on dermal mast cells provokes release and synthesis of a vast series of mediators. Following their recruitment and activation into the skin, eosinophils are also thought to contribute to tissue damage. Keratinocytes of AD patients exhibit a propensity for exaggerated production of cytokines and chemokines, a phenomenon that can be relevant in promoting and maintaining inflammation, and may have a major role in localizing the atopic diathesis to the skin (Girolomoni and Pastore 2001). Thus, a complex network of cytokines and chemokines contributes to establish a local milieu that favors the permanence of inflammation in AD skin. Patients with AD exhibit exaggerated Th2 responses, and initiation of AD lesions is thought to be mediated by early skin infiltration of Th2 lymphocytes releasing high levels of IL-4, IL-5, and IL-13. Subsequently, the accumulation of activated monocytes, mature DCs, and eosinophils determines a rise in IL-12 expression and the appearance of a mixed Th2/Th1 cytokine pattern, with reduced IL-4 and IL-13 and the presence of IFN-γ.

The proportion of CD4$^+$ T lymphocytes expressing the CCR4 receptor in the peripheral blood of patients with AD is higher compared to CD4$^+$ T cells of healthy controls. In contrast, AD patients bear a lower percentage of circulating CXCR3$^+$CD4$^+$ T cells (Vestergaard et al. 2000; Yamamoto et al. 2000; Nakatani et al. 2001; Wakugawa et al. 2001). Moreover, the percentage of blood CCR4$^+$CD4$^+$ cells correlates positively with disease severity and IL-4 and IL-13 secretion by CD4$^+$ T cells (Wakugawa et al. 2001; Okazaki et al. 2002). CCR4$^+$CD4$^+$ T cells are also positive for the skin-homing receptor, CLA, and infiltrate AD lesions in high numbers (Nakatani et al. 2001; Wakugawa et al. 2001), indicating not only increased generation of CCR4$^+$ T cells, but also enhanced recruitment into AD skin. The ligands for CCR4 are TARC and MDC, two chemokines present in high amounts in the plasma of AD patients and whose levels also correlate with disease activity (Fujisawa et al. 2002; Horikawa et al. 2002; Kakinuma et al. 2002). Both TARC and MDC are produced abundantly by immature DCs and even more by mature DCs in vitro. In addition, in situ studies have shown that MDC immunoreactivity in AD skin is mostly confined to CD1a$^+$CD83$^+$ mature DCs, and identify this cell type as the major source of MDC in vivo (Vulcano et al. 2001; D'Ambrosio et al. 2002). In aggregate, these data point out that CCR4$^+$ T cells may be relevantly involved in the pathogenesis of AD, and that DCs may guide not only their activation but also their preferential accumulation in AD skin. TARC is also expressed on microvascular endothelial cells in AD lesions, and thus may be primarily involved in the arrest of CCR4$^+$ T cells (D'Ambrosio et al. 2002). Other chemokines that participate in the accumulation of T cells in AD include RANTES, MCP-1, eotaxin, MIP-3a, CTACK, and IL-16. RANTES and MCP-1, which attract both Th1 and Th2 cells, are expressed by infiltrating leukocytes but especially keratinocytes in diseased skin (Giustizieri et al. 2001), although only RANTES is elevated in the serum of patients (Kaburagi et al. 2001). Eotaxin, together with its receptor CCR3, is expressed in the dermis of AD lesions, particularly by mononuclear cells and fibroblasts (Yawalkar et al. 1999; Taha et al. 2000). MIP-3a mRNA is expressed in AD skin, although less abundantly than in psoriasis (Schmuth et al. 2002), with immunostaining localizing the chemokine in the basal epidermis, and CCR6$^+$ cells being mainly DCs and T cells (Nakayama et al. 2000). Interestingly, disruption of the epidermal permeability barrier upregulates epidermal MIP-3a mRNA, revealing an important mechanism for the initial influx of DCs and T cells in AD skin, which constitutively presents epidermal permeability barrier dysfunction (Schmuth et al. 2002). Similarly to psoriasis, acute and chronic AD lesions exhibit strong CTACK expression in the epidermis and numerous CCR10$^+$ T cells (Homey et al. 2002). Recently, an elevation of circulating IL-16 has been associated to active AD in children (Frezzolini et al. 2002). Increased expression of IL-16 has already been detected in the epidermis of AD lesions (Laberge S et al. 2001), and Langerhans cells (LCs) have been recognized as the most relevant source of this chemokine in this disease (Reich et al. 2002). IL-16 is a strong inducer of migratory responses in CD4$^+$ T cells, monocytes, and eosinophils (Cruikshank et al. 2000). In vitro experiments indicate that IL-16 is a major chemotactic signal from DCs toward themselves and CD4$^+$ T cells (Kaser et al. 1999). It is thus possible to speculate that LC-derived IL-16 may critically contribute to T cell recruitment in AD. Indeed, lesional skin of AD patients exhibits an increased number of cells belonging to the DC lineage, including epidermal LCs, dermal DCs, and a unique population of epidermal CD1a$^+$ DCs expressing CD1b and/or CD36, which closely resemble DCs generated in vitro by culturing monocytes in the presence of GM-CSF and IL-4. Such DCs can efficiently present IgE-bound allergens to T lymphocytes, since they display an upregulated expression of both the high affinity (FcεRI) as well as the low affinity (FcεRII/CD23) IgE receptors (Stingl and Maurer 1997). Furthermore, FcεRI engagement has been shown to upregulate IL-16 production in LCs derived from atopic donors (Reich et al. 2001b). Eventually, the selective uptake of allergens and the subsequent induction of specific T-cell responses may further provide a mechanism to perpetuate skin inflammation.

LC-derived IL-16 can also attract eosinophils into the skin. Although eosinophils are not prominent in the AD infiltrate, they are potent effector cells, and contribute to inflammation by the release of a variety of cytotoxic species such as eosinophil cationic protein, eosinophil peroxidase major basic protein, and eosinophil-derived neurotoxin/eosinophil protein X. In the context of acute and chronic AD lesions, eosinophils are attracted by resident skin populations through the increased expression of CCR3 binding molecules, including RANTES, MCP-4, and eotaxin (Yawalkar et al. 1999; Taha et al. 2000). In situ hybridization experiments performed on skin biopsies soon after challenge with a proper provocation factor have demonstrated a prominent neosynthesis of RANTES and MCP-3 by dermal fibroblasts. In acute AD, eosinophil attraction could be predominantly performed by these cells, whose eotaxin and MCP-4 secretion is potently increased in response to T cell-derived IL-4 (Mochizuki et al. 1998; Petering et al. 1998).

In contrast to psoriasis, IL-8 and IP-10 are only weakly expressed in some limited areas of the epidermis in AD lesions. In vitro studies have shown that keratinocytes from AD patients produced increased amounts of RANTES, but reduced levels of IP-10, in response to IFN-γ or TNF-α when compared to keratinocytes from normal controls or psoriasis patients (Giustizieri et al. 2001). RANTES is not the only proinflammatory factor whose expression has been found upregulated in keratinocytes cultured from AD patients, since they displayed overproduction of spontaneous as well as IL-1α- and IFN-γ-induced GM-CSF release, when compared to healthy control keratinocytes (Pastore et al. 1997). Numerous functional polymorphisms in the regulatory/coding regions of clusters of cytokine/chemokine genes, including RANTES, have been found in AD patients (Nickel et al. 2000; Elliot and Forrest 2002), which could be implicated in overproduction by keratinocytes. However, apart from genes coding for Th2 cytokines, polymorphisms for other inflammatory genes have not been confirmed in other studies (Kozma et al. 2002). Indeed, the genes that contribute to complex diseases are difficult to identify because they typically exert small effects on disease risk; in addition, the magnitude of their effects is likely to be modified by other unrelated genes as well as environmental factors. Thus, susceptibility loci for complex diseases identified in one study may not be replicated in other populations. In searching for a molecular mechanism underlying abnormal GM-CSF production, we have found that AD keratinocytes express higher constitutive or induced levels of members of the AP-1 family of transcription factors, including c-Jun, JunB, and c-Fos, compared to keratinocytes from normal controls (Pastore et al. 2000). AP-1 is prominently activated by various cytokines, including IL-4, IFN-γ, and TNF-α, and AP-1 binding sites are strategically located in the promoters of a vast array of cytokines and chemokines, including RANTES. In conclusion, our data support the hypothesis that contribution of keratinocytes to the pathogenesis of AD and psoriasis is linked to the presence of distinct, intrinsic alterations in their capacity to respond to proinflammatory stimuli, and that these abnormalities can be important in the inflammatory hyperreactivity of AD skin. In particular, epithelial/DC interactions may be very important in the initiation and persistence of inflammation in AD. The propensity of AD keratinocytes to produce higher than normal levels of growth factors (GM-CSF), chemokines (RANTES), and cytokines (TNF-α, thymic stromal lymphopoietin), may greatly stimulate DC differentiation from precursors, and recruit as well as activate DCs in AD skin. This activation includes high production of chemokines attracting CCR4$^+$ Th2 lymphocytes and increased stimulation of T cell responses (Soumelis et al. 2002). The triggers that activate keratinocytes in the very early phases may include the altered epidermal permeability barrier functions (Elias et al. 1999; Schmuth et al. 2002). In contrast to bronchial epithelial cells, environmental allergens such as those of the house dust mite do not seem to stimulate keratinocyte production of chemokines or cytokines (Mascia et al. 2002).

11.8 Concluding Remarks

Chemokines appear to be crucial regulators of both the induction and expression of chronic inflammatory skin diseases. ACD is serving as a valuable model for understanding the specific contribution

Table 3. Chemokine expression in inflammatory skin diseases

	Major cell sources	Allergic contact dermatitis	Psoriasis	Atopic dermatitis
IL-8	Keratinocytes Mast cells	++	++	+
IP-10	Keratinocytes T cells Monocytes DCs	+++	++	+
I-309	Keratinocytes T cells Monocytes	+	nd	nd
Eotaxin	Fibroblasts T cells Mast cells	nd	nd	++
MCP-1	Keratinocytes T cells, mast cells Monocytes DCs	++	++	++
MCP-3, MCP-4 RANTES	Fibroblasts Keratinocytes T cells Monocytes DCs	+ ++	+ ++	+++ ++
MDC	DCs	+	+	+++
TARC	Dendritic cells Endothelial cells	+	+	+
CTACK	Keratinocytes	++	++	++
MIP-3a	Keratinocytes Endothelial cells	++	+++	nd
Fractalkine	Keratinocytes Endothelial cells	++	++	+

nd, not determined.

of different T cell subsets as well as the mechanisms underlying the generation and regulation of T cell responses. The kinetics and pattern of chemokine expression during ACD resembles those observed during wound healing, with IL-8 and MCP-1 expressed first, followed by RANTES, and finally by CXCR3 agonists (Engelhardt et al. 1998; Goebeler et al. 2001), suggesting that the skin sets up a standard sequential pattern of chemokine expression in response to different types of injuries. However, psoriasis and atopic dermatitis differ substantially in the pattern of chemokines expressed in the skin (Table 3), and the level of certain chemokines may differentiate ACD from irritant contact dermatitis (Flier et al. 1999).

Several companies have reported the development of selective small-molecule chemokine receptor antagonists, and some of these compounds have entered phase I clinical trials (Carter 2002). Some evidence suggests that chemokine receptor antagonism is a reasonable therapeutic strategy for inflammation. However, given the promiscuity of the chemokine–chemokine receptor axis, it is difficult to foresee whether blocking one chemokine or receptor may be sufficient to counteract skin inflammatory responses. In this context, it is worth mentioning the clinical failure of platelet-activating factor antagonism in the systemic inflammatory response syndrome associated with acute pancreatitis (Johnson et al 2001). The use of multiple selective chemokine receptor antagonists may provide a more efficacious approach. Indeed, promising results have been obtained with dual-receptor antagonists, such as Met-RANTES, which antagonizes the actions of RANTES on both CCR1 and CCR5, and dramatically reduces tissue damage in a colitis model, eosinophilia in a pulmonary inflammation model, and disease symptoms in collagen-induced arthritis (Ajuebor et al 2001; Gonzalo et al 1998). Most of the studies on the role of chemokines in CH have looked at the induction phase and few if any have addressed the expression phase. However, work in other models of Th1-mediated inflammation has provided interesting results (Fife et al. 2001; Salomon et al. 2002). The growing knowledge of the mechanisms controlling the recruitment of T cells in the skin during inflammatory and immune responses will certainly permit the design of novel therapeutic approaches targeting chemokines or their receptors, with the possibility of effectively blocking disease activity.

References

Abe M, Kondo T, Xu H, Fairchild RL (1996) Interferon-γ inducible protein (IP-10) expression is mediated by CD8$^+$ T cells and is regulated by CD4$^+$ T cells during the elicitation phase of contact hypersensitivity. J Invest Dermatol 107:360–366

Ajuebor MN, Hogaboam CM, Kunkel SL, Proudfoot AE, Wallace JL (2001) The chemokine RANTES is a crucial mediator of the progression from acute to chronic colitis in the rat. J Immunol 166:552–558

Albanesi C, Cavani A, Girolomoni G (1999) Interleukin-17 is produced by nickel-specific T lymphocytes and regulates ICAM-1 expression and chemokine production in human keratinocytes. Synergistic or antagonist effects with interferon-γ and tumor necrosis factor-α. J Immunol 162:494–502

Albanesi C, Scarponi C, Federici M, Nasorri F, Cavani A, Girolomoni G (2000a) Interleukin 17 is produced by both Th1 and Th2 lymphocytes, and modulates interferon-γ- and interleukin 4-induced activation of human keratinocytes. J Invest Dermatol 115:81–87

Albanesi C, Scarponi C, Sebastiani S, Cavani A, Federici M, De Pità O, Puddu P, Girolomoni G (2000b) IL-4 enhances keratinocyte expression of CXCR3 agonistic chemokines. J Immunol 165:1395–1402

Albanesi C, Scarponi C, Sebastiani S, Cavani A, Federici M, Sozzani S, Girolomoni G (2001) A cytokine-to-chemokine axis between T lymphocytes and keratinocytes can favor Th1 cell accumulation in chronic inflammatory skin diseases. J Leukoc Biol 70:617–623

Anttila HS, Reitamo S, Erkko P, Ceska M, Moser B, Baggiolini M (1992) Interleukin-8 immunoreactivity in the skin of healthy subjects and patients with palmoplantar pustulosis and psoriasis. J Invest Dermatol 98:96–101

Asadullah K, Friedrich M, Hanneken S, Rohrbach C, Audring H, Vergopoulos A, Ebeling M, Docke WD, Volk HD, Sterry W (2001) Effects of systemic interleukin-10 therapy on psoriatic skin lesions: histologic, immunohistologic, and molecular biology findings. J Invest Dermatol 116:721–727

Asadullah K, Volk H-D, Sterry W (2002) Novel immunotherapies for psoriasis. Trends Immunol 23:47–53

Bazan JF, Bacon KB, Hardiman G, Wang W, Soo K, Rossi D, Greaves DR, Zlotnik A, Schall TJ (1997) A new class of membrane-bound chemokine with a CX3C motif. Nature 385:640–644

Biedermann T, Kneilling M, Mailhammer R, Maier K, Sander CA, Kollias G, Kunkel SL, Hültner L, Röcken M (2000) Mast cells control neutrophil recruitment during T cell-mediated delayed-type hypersensitivity reactions through tumor necrosis factor and macrophage inflammatory protein 2. J Exp Med 192:1441–1451

Biedermann T, Mailhammer R, Mai A, Sander C, Ogilvie A, Brombacher F, Maier K, Levine AD, Röcken M (2001) Reversal of established delayed type hypersensitivity reactions following therapy with IL-4 or antigen-specific Th2 cells. Eur J Immunol 31:1582–1591

Bodgan C (2001) Nitric oxide and the immune response. Nat Immunol 2:907–916

Bonecchi R, Bianchi G, Bordignon PP, D'Ambrosio D, Lang R, Borsatti A, Sozzani S, Allavena P, Gray PA, Mantovani A, Sinigaglia F (1998) Differential expression of chemokine receptors and chemotactic responsiveness of type 1 T helper cells (Th1 s) and Th2 s. J Exp Med 187:129–134

Bouloc A, Cavani A, Katz SI (1998) Contact hypersensitivity in MHC class II-deficient mice depends on CD8$^+$ T lymphocytes primed by immunostimulating Langerhans cells. J Invest Dermatol 111:44–49

Campbell JJ, Haraldsen G, Pan J, Rottman J, Qin S, Ponath P, Andrew DP, Warnke R, Ruffing N, Kassam N, Wu L, Butcher EC (1999) The chemokine receptor CCR4 in vascular recognition by cutaneous but not intestinal memory T cells. Nature 400:776–780

Carter PH (2002) Chemokine receptor antagonism as a approach to anti-inflammatory therapy: "just right" or plain wrong? Curr Opin Chem Biol 6:510–525

Cavani A, Nasorri F, Prezzi C, Sebastiani S, Albanesi C, Girolomoni G (2000) Human CD4$^+$ T lymphocytes with remarkable regulatory functions on dendritic cells and nickel-specific Th1 immune responses. J Invest Dermatol 114:295–302

Cavani A, Albanesi C, Traidl C, Sebastiani S, Girolomoni G (2001) Effector and regulatory T cells in allergic contact dermatitis. Trends Immunol 22:118–120

Charbonnier A-S, Kohrgruber N, Kriehuber E, Stingl G, Rot A, Maurer D (1999) Macrophage inflammatory protein 3α is involved in the constitutive trafficking of epidermal Langerhans cells. J Exp Med 190:1755–1767

Corinti S, Chiarantini L, Dominici S, Laguardia ME, Magnani M, Girolomoni G (2002) Erythrocytes deliver Tat to interferon-γ-treated human dendritic cells for efficient initiation of specific type 1 immune responses in vitro. J Leukoc Biol 71:652–658

Cruikshank WW, Kornfeld H, Center DM (2000) Interleukin-16. J Leukoc Biol 67:757–766

D'Ambrosio D, Iellem A, Bonecchi R, Mazzeo D, Sozzani S, Mantovani A, Sinigaglia F (1998) Selective upregulation of chemokine receptors CCR4 and CCR8 upon activation of polarized human type 2 Th cells. J Immunol 161:5111–5115

D'Ambrosio D, Albanesi A, Lang R, Girolomoni G, Sinigaglia F, Laudanna C (2002) Quantitative differences in chemokine receptor engagement generate diversity in integrin-dependent lymphocyte adhesion. J Immunol 169:2303–2312

Dieu-Nosejan M-C, Massacier C, Homey B, Vanbervliet B, Pin J-J, Vicari A, Lebecque S, Dezutter-Dambuyant C, Schmitt D, Zlotnik A, Caux C

(2000) Macrophage inflammatory protein 3α is expressed at inflamed epithelial cell surfaces and is the most potent chemokine known in attracting Langerhans cell precursors. J Exp Med 192:705–717

Dulkys Y, Schramm G, Kimmig D, Knoss S, Weyergraf A, Kapp A, Elsner J (2001) Detection of mRNA for eotaxin-2 and eotaxin-3 in human dermal fibroblasts and their distinct activation profile on human eosinophils. J Invest Dermatol 116:498–505

Elias PM, Wood LC, Feingold KR (1999) Epidermal pathogenesis of inflammatory dermatoses. Am J Contact Dermat 10:119–126

Elliot K, Forrest S (2002) Genetics of atopic dermatitis. In: Bieber T, Leung DYM (eds) Atopic dermatitis. Marcel Dekker, New York, pp 81–110

Engelhardt E, Toksoy A, Goebeler M, Debus S, Bröcker E-B, Gillitzer R (1998) Chemokines IL-8, Gro-α, MCP-1, IP-10, and MIG are sequentially and differentially expressed during phase-specific infiltration of leukocyte subset in human wound healing. Am J Pathol 153:1849–1860

Federici M, Giustizieri ML, Scarponi C, Girolomoni G, Albanesi C (2002) Impaired IFN-γ-dependent inflammatory responses in human keratinocytes overexpressing the suppressor of cytokine signaling 1. J Immunol 169:434–442

Fife BT, Kennedy KJ, Paniagua MC, Lukacs NW, Kunkel SL, Luster AD, Karpus WJ (2001) CXCL10 (IFN-γ-inducible protein-10) control of encephalitogenic CD4$^+$ T cell accumulation in the central nervous system during experimental autoimmune encephalomyelitis. J Immunol 166:7617–7624

Fitzhugh DJ, Naik S, Caughman SW, Hwang ST (2000) C-C chemokine receptor 6 is essential for arrest of a subset of memory T cells on activated dermal microvascular endothelial cells under physiologic flow conditions in vitro. J Immunol 165:6677–6681

Flier J, Boorsma DM, Bruynzeel DP, van Beek PJ, Stoof TJ, Scheper RJ, Willemze R, Tensen CP (1999) The CXCR3 activating chemokines IP-10, Mig, and IP-9 are expressed in allergic but not in irritant patch test reactions. J Invest Dermatol 113:574–578

Förster R, Schubel A, Breitfeld D, Kremmer E, Renner-Müller I, Wolf E, Lipp M (1999) CCR7 coordinates the primary immune response by establishing functional microenvironments in secondary lymphoid organs. Cell 99:23–33

Fraticelli P, Sironi M, Bianchi G, D'Ambrosio D, Albanesi C, Stopacciaro A, Chieppa M, Allavena P, Ruco L, Girolomoni G, Sinigaglia F, Vecchi A, Mantovani A (2001) Fractalkine (CX3CL1) as an amplification circuit of polarized Th1 responses. J Clin Invest 107:1173–1181

Frezzolini A, Paradisi M, Zaffiro A, Provini A, Cadoni S, Ruffelli M, De Pita O (2002) Circulating interleukin 16 (IL-16) in children with atopic/eczema dermatitis syndrome (AEDS): a novel serological marker of disease activity. Allergy 57:815–820

Fujisawa T, Fujisawa R, Kato Y, Nakayama T, Morita A, Katsumata H, Nishimori H, Iguchi K, Kamiya H, Gray PW, Chantry D, Suzuki R,

Yoshie O (2002) Presence of high contents of thymus and activation-regulated chemokine in platelets and elevated plasma levels of thymus and activation-regulated chemokine and macrophage-derived chemokine in patients with atopic dermatitis. J Allergy Clin Immunol 110:139–146

Gillitzer R, Wolff K, Tong D, Müller C, Yoshimura T, Hartmann AA, Stingl G, Berger R (1993) MCP-1 mRNA expression in basal keratinocytes of psoriatic lesions. J Invest Dermatol 101:127–131

Girolomoni G, Sebastiani S, Albanesi C, Cavani A (2001) T-cell subpopulations in the development of atopic and contact allergy. Curr Opin Immunol 13:733–737

Girolomoni G, Pastore S (2001) Epithelial cells and atopic diseases. Curr Allergy Asthma Rep 1:481–482

Giustizieri ML, Albanesi C, Scarponi C, De Pità O, Girolomoni G (2002) Nitric oxide donors suppress chemokine production by keratinocytes in vitro and in vivo. Am J Pathol 161:1409–1418

Giustizieri ML, Mascia F, Frezzolini A, De Pità O, Chinni ML, Giannetti A, Girolomoni G, Pastore S (2001) Keratinocytes from patients with atopic dermatitis and psoriasis show a different chemokine production profile in response to T cell-derived cytokines. J Allergy Clin Immunol 107:871–877

Goebeler M, Trautmann A, Voss A, Bröcker E-B, Toksoy A, Gillitzer R (2001) Differential and sequential expression of multiple chemokines during elicitation of allergic contact hypersensitivity. Am J Pathol 158:431–440

Gonzalo JA, Lloyd CM, Wen D, Albar JP, Wells TN, Proudfoot A, Martinez AC, Dorf M, Bjerke T, Coyle AJ, Gutierrez-Ramos JC (1998) The coordinated action of CC chemokines in the lung orchestrates allergic inflammation and airway hyperresponsiveness. J Exp Med 188:157–167

Gottlieb AB, Luster AD, Posnett DN, Carter DM (1998) Detection of a γ interferon-induced protein IP-10 in psoriatic plaques. J Exp Med 168:941–948

Griffiths CEM, Barker JN, Kunkel S, Nickoloff BJ (1991) Modulation of leukocyte adhesion molecule, a T-cell chemotaxin (IL-8) and a regulatory cytokine (TNFa) in allergic contact dermatitis (rhus dermatitis). Br J Dermatol 124:519–531.

Gunn MD, Tangemann K, Tam C, Cyster JG, Rosen SD, Williams LT (1998) A chemokine expressed in lymphoid high endothelial venules promotes the adhesion and chemotaxis of naive T lymphocytes. Proc Natl Acad Sci USA 95:258–263

Gunn MD, Kyuwa S, Tam C, Kakiuchi T, Matsukawa A, Williams LT, Nakano H (1999) Mice lacking expression of secondary lymphoid organ chemokine have defects in lymphocyte homing and dendritic cell localization. J Exp Med 189:451–460

Haase I, Hobbs RM, Romero MR, Broad S, Watt FM (2001) A role for mitogen-activated protein-kinase activation by integrins in the pathogenesis of psoriasis. J Clin Invest 108:527–536

Homey B, Dieu-Nosejan M-C, Wiesenborn A, Massacrier C, Pin J-J, Oldham E, Catron D, Buchanan ME, Müller A, deWaal Malefyt R, Deng G, Orozco R, Ruzicka T, Lehman P, Lebecque S, Caux C, Zlotnik A (2000a) Upregulation of macrophage inflammatory protein 3a/CCL20 and CC chemokine receptor 6 in psoriasis. J Immunol 164:6621–6632

Homey B, Wang W, Soto H, Buchanan ME, Wiesenborn A, Catron D, Muller A, McClanahan TK, Dieu-Nosjean MC, Orozco R, Ruzicka T, Lehmann P, Oldham E, Zlotnik A (2000b) The orphan chemokine receptor G protein-coupled receptor-2 (GPR-2, CCR10) binds the skin-associated chemokine CCL27 (CTACK/ALP/ILC). J Immunol 164:3465–3470

Homey B, Alenius H, Muller A, Soto H, Bowman EP, Yuan W, McEvoy L, Lauerma AI, Assmann T, Bunemann E, Lehto M, Wolff H, Yen D, Marxhausen H, To W, Sedgwick J, Ruzicka T, Lehmann P, Zlotnik A (2002) CCL27-CCR10 interactions regulate T cell-mediated skin inflammation. Nat Med 8:157–165

Horikawa T, Nakayama T, Hikita I, Yamada H, Fujisawa R, Bito T, Harada S, Fukunaga A, Chantry D, Gray PW, Morita A, Suzuki R, Tezuka T, Ichihashi M, Yoshie O (2002) IFN-γ-inducible expression of thymus and activation-regulated chemokine/CCL17 and macrophage-derived chemokine/CCL22 in epidermal keratinocytes and their roles in atopic dermatitis. Int Immunol 14:767–773

Hudak S, Hagen M, Liu Y, Catron D, Oldham E, McEvoy LM, Bowman EP (2002) Immune surveillance and effector functions of CCR10$^+$ skin homing T cells. J Immunol 169:1189–1196Iellem A, Mariani M, Lang R, Recalde H, Panina-Bordignon P, Sinigaglia F, D'Ambrosio D (2001) Unique chemotactic response profile and specific expression of chemokine receptors CCR4 and CCR8 by CD4$^+$CD25$^+$ regulatory T cells. J Exp Med 194:847–853

Johnson CD, Kingsnorth AN, Imne CW, McMahon MJ, Neoptolemos JP, McKay C, Toh SK, Skaife P, Leeder PC, Wilson P, Larvin M, Curtis LD (2001) Double blind, randomised, placebo controlled study of a platelet activating factor antagonist, lexipafant, in the treatment and prevention of organ failure in predicted severe acute pancreatitis. Gut 48:62–69

Kaburagi Y, Shimada Y, Nagaoka T, Hasegawa M, Takehara K, Sato S (2001) Enhanced production of CC-chemokines (RANTES, MCP-1, MIP-1α, MIP-1β, and eotaxin) in patients with atopic dermatitis. Arch Dermatol Res 293:350–355

Kakinuma T, Nakamura K, Wakugawa M, Mitsui H, Tada Y, Saeki H, Torii H, Komine M, Asahina A, Tamaki K (2002) Serum macrophage-derived chemokine (MDC) levels are closely related with the disease activity of atopic dermatitis. Clin Exp Immunol 27:270–273

Karvonen SL, Korkiamaki T, Yla-Outinen H, Nissinen M, Teerikangas H, Pummi K, Karvonen J, Peltonen J (2000) Psoriasis and altered calcium metabolism: downregulated capacitative calcium influx and defective calcium-mediated cell signaling in cultured psoriatic keratinocytes. J Invest Dermatol 114:693–700

Kaser A, Dunzendorfer S, Offner FA, Ryan T, Schwabegger A, Cruikshank WW, Wiedermann CJ, Tilg H (1999) A role for IL-16 in the cross-talk between dendritic cells and T cells. J Immunol 163:3232–3238

Kehren J, Desvignes C, Krasteva M, Ducluzeau M-T, Assossou O, Horand F, Hahne M, Kägi D, Kaiserlian D, Nicolas J-F (1999) Cytotoxicity is mandatory for CD8$^+$ T cell-mediated contact hypersensitivity. J Exp Med 189:779–786

Koelle DM, Liu Z, McClurkan CM, Topp MS, Riddell SR, Pamer EG, Johnson AS, Wald A, Corey L (2002) Expression of cutaneous lymphocyte-associated antigen by CD8$^+$ T cells specific for a skin-tropic virus. J Clin Invest 110:537–548

Kozma GT, Falus A, Bojszko A, Krikovszky D, Szabo T, Nagy A, Szalai C (2002) Lack of association between atopic eczema/dermatitis syndrome and polymorphisms in the promoter region of RANTES and regulatory region of MCP-1. Allergy 57:160–163

Kraft S, Novak N, Katoh N, Bieber T, Rupec RA (2002) Aggregation of the high-affinity IgE receptor FcεRI on human monocytes and dendritic cells induces NF-κB activation. J Invest Dermatol 118:830–837

Krueger JG (2002) The immunologic basis for the treatment of psoriasis with new biological agents. J Am Acad Dermatol 46:1–23

Kulke R, Bornscheuer E, Schluter C, Bartels J, Rowert J, Sticherling M, Christophers E (1998) The CXC receptor 2 is overexpressed in psoriatic epidermis. J Invest Dermatol 110:90–94

Kunkel EJ, Butcher EC (2002) Chemokines and the tissue-specific migration of lymphocytes. Immunity 16:1–4

Laberge S, Ghaffar O, Boguniewicz M, Center DM, Leung DY, Hamid Q (1998) Association of increased CD4+ T-cell infiltration with increased IL-16 expression in atopic dermatitis. J Allergy Clin Immunol 102:645–650

la Sala A, Sebastiani S, Ferrari D, Di Virgilio F, Izdko M, Norgauer J, Girolomoni G (2002) Dendritic cells exposed to extracellular adenosine triphosphate acquire the migratory properties of mature cells and show a reduced capacity to attract type 1 T lymphocytes. Blood 99:1715–1722

Leung DY, Soter NA (2001) Cellular and immunologic mechanisms in atopic dermatitis. J Am Acad Dermatol 44:S1–S12

Liao F, Rabin RL, Smith CS, Sharma G, Nutman TB, Farber JM. (1999) CC-chemokine receptor 6 is expressed in diverse memory subsets of T cells and determines responsiveness to macrophage inflammatory protein 3a. J Immunol 162:186–194

Loetscher P, Pellegrino A, Gong J-H, Mattioli, I, Loetscher M, Bardi G, Baggiolini M, Clark-Lewis I (2001) The ligands of CXC receptor 3, I-TAC, Mig, and IP-10 are natural antagonists for CCR3. J Biol Chem 276:2986–2991

Luster AD (2002) The role of chemokines in linking innate and adaptive immunity. Curr Opin Immunol 14:129–135

Luther SA, Cyster JG (2001) Chemokines as regulators of T cell differentiation. Nat Immunol 2:102–107

Mackay CR (2001) Chemokines: immunology's high impact factors. Nat Immunol 2:95–101

Mascia F, Mariani V, Giannetti A, Girolomoni G, Pastore S (2002) House dust mite allergen exerts no direct proinflammatory effects on human keratinocytes. J Allergy Clin Immunol 109:532–538

Mekory YA, Metcalfe DD. (1999) Mast cell-T cell interactions. J Allergy Clin Immunol 104:517–523

Mellado M, Rodríguez-Frade JM, Mañes S, Martinez-A C (2001) Chemokine signaling and functional responses: the role of receptor dimerization and TK pathway activation. Annu Rev Immunol 19:397–421

Mochizuki M, Bartels J, Mallet AI, Christophers E, Schröder JE (1998) IL-4 induces eotaxin: a possible mechanism of selective eosinophils recruitment in helminth infection and atopy. J Immunol 160:60–68

Morales J, Homey B, Vicari AP, Hudak S, Oldham E, Hedrick J, Orozco R, Copeland NG, Jenkins NA, McEvoy LM, Zlotnik A (1999) CTACK, a skin-associated chemokine that preferentially attracts skin-homing memory T cells. Proc Natl Acad Sci USA 96:14470–14475

Moser B, Loetscher P (2001) Lymphocyte traffic control by chemokines. Nat Immunol 2:123–128

Nakamura K, Williams IR, Kupper TS (1995) Keratinocyte-derived monocyte chemoattractant protein 1 (MCP-1): analysis in a transgenic model demonstrates MCP-1 can recruit dendritic and Langerhans cells to skin. J Invest Dermatol 105:635–643

Nakatani T, Kaburagi Y, Shimada Y, Inaoki M, Takehara K, Mukaida N, Sato S (2001) CCR4+ memory CD4+ T lymphocytes are increased in peripheral blood and lesional skin from patients with atopic dermatitis. J Allergy Clin Immunol 107:353–358

Nakayama N, Fujisawa R, Yamada H, Horikawa T, Kawasaki H, Hieshima K, Izawa D, Fujiie S, Tezuka T, Yoshie O (2000) Inducible expression of a CC chemokine liver- and activation-regulated chemokine (LARC)/macrophage inflammatory protein (MIP)-3a/CCL20 by epidermal keratinocytes and its role in atopic dermatitis. Int Immunol 13:95–103

Nickel RG, Casolaro V, Wahn U, Beyer K, Barnes KC, Plunkett BS, Freidhoff LR, Sengler C, Plitt JR, Schleimer RP, Caraballo L, Naidu RP, Levett PN, Beaty TH, Huang SK (2000) Atopic dermatitis is associated with a functional mutation in the promoter of the C-C chemokine RANTES. J Immunol 164:1612–1616.

Nickoloff BJ, Mitra RS, Varani J, Dixit VM, Polverini PJ (1994) Aberrant production of interleukin-8 and thrombospondin-1 by psoriatic keratinocytes mediates angiogenesis. Am J Pathol 144:820–828

Nickoloff BJ, Schröder JM, von den Driesch P, Raychaudhuri SP, Farber EM, Boehncke WH, Morhenn VB, Rosenberg EW, Schön MP, Holick MF (2000) Is psoriasis a T-cell disease? Exp Dermatol 9:359–375

Ogilvie P, Bardi G, Clark-Lewis I, Baggiolini M, Uguccioni M (2001) Eotaxin is a natural antagonist for CCR2 and an agonist for CCR5 Blood 97:1920–1924

Okazaki H, Kakurai M, Hirata D, Sato H, Kamimura T, Onai N, Matsushima K, Nakagawa H, Kano S, Minota S (2002) Characterization of chemokine receptor expression and cytokine production in circulating CD4$^+$ T cells from patients with atopic dermatitis: upregulation of C-C chemokine receptor 4 in atopic dermatitis. Clin Exp Allergy 32:1236–1242

Ormerod AD, Weller R, Copeland P, Benjamin N, Ralston SH, Grabowksi P, Herriot R (1998) Detection of nitric oxide and nitric oxide synthases in psoriasis. Arch Dermatol Res 290:3–8

Pastore S, Fanales-Belasio E, Albanesi C, Chinni ML, Giannetti A, Girolomoni G (1997) Granulocyte macrophage colony-stimulating factor is overproduced by keratinocytes in atopic dermatitis. Implications for sustained dendritic cell activation in the skin. J Clin Invest 99:3009–3017

Pastore S, Giustizieri M, Mascia F, Giannetti A, Kaushansky K, Girolomoni G (2000) Dysregulated activation of activator protein 1 in keratinocytes of atopic dermatitis patients with enhanced expression of granulocyte/macrophage-colony stimulating factor. J Invest Dermatol 115:1134–1143

Petering H, Höchstetter R, Kimming D, Smolarski R, Kapp A, Elsner J (1998) Detection of MCP-4 in dermal fibroblast and its activation of the respiratory burst in human eosinophils. J Immunol 160:555–558

Reich K, Garbe C, Blaschke V, Maurer C, Middel P, Westphal G, Lippert U, Neumann C (2001a) Response of psoriasis to interleukin-10 is associated with suppression of cutaneous type 1 inflammation, downregulation of the epidermal interleukin-8/CXCR2 pathway and normalization of keratinocyte maturation. J Invest Dermatol 116:319–329

Reich K, Heine A, Hugo S, Blaschke V, Middel P, Kaser A, Tilg H, Blaschke S, Gutgesell C, Neumann C (2001b) Engagement of FcεRI stimulates the production of IL-16 in Langerhans cell-like dendritic cells. J Immunol 167:6321–6329

Reich K, Hugo S, Middel P, Blaschke V, Heine A, Gutgesell C, Williams R, Neumann C (2002) Evidence for a role of Langerhans cell-derived IL-16 in atopic dermatitis. J Allergy Clin Immunol 109:681–687

Rennert PD, Hochman PS, Flavell RA, Chaplin DD, Jayaraman S, Browning JL, Fu Y-X (2001) Essential role of lymph nodes in contact hypersensitivity revealed in lymphotoxin-α-deficient mice. J Exp Med 193:1227–1238

Robert C, Kupper TS (1999) Inflammatory skin diseases, T cells, and immune surveillance. N Engl J Med 341:1817–1828

Rossi D, Zlotnik A (2000) The biology of chemokines and their receptors. Annu Rev Immunol 18:217–242

Rottman JB, Smith TL, Ganley KG, Kikuchi T, Krueger JG (2001) Potential role of the chemokine receptors CXCR3, CCR4, and the integrin αEβ7 in the pathogenesis of psoriasis vulgaris. Lab Invest 8:335–347

Sallusto F, Kremmer E, Palermo B, Hoy A, Ponath P, Qin S, Forster R, Lipp M, Lanzavecchia A (1999a) Switch in chemokine receptor expression upon TCR stimulation reveals novel homing potential for recently activated T cells. Eur J Immunol 29:2037–2045

Sallusto F, Lenig D, Forster R, Lipp M, Lanzavecchia A (1999b) Two subsets of memory T lymphocytes with distinct homing potential and effector function. Nature 401:708–712

Sallusto F, Palermo B, Lenig D, Miettinen M, Matikainen S, Julkunen I, Forster R, Brugstahler R, Lipp M, Lanzavecchia A (1999c) Distinct patterns and kinetics of chemokine production regulate dendritic cell function. Eur J Immunol 29:1617–1625

Sallusto F, Mackay CF, Lanzavecchia A (2000) The role of chemokine receptors in primary, effector, and memory immune responses. Annu Rev Immunol 18:593–620

Salomon I, Netzer N, Wildbaum G, Schif-Zuck S, Maor G, Karin N (2002) Targeting the function of IFN-γ-inducible Protein 10 suppresses ongoing adjuvant arthritis. J Immunol 169:2685–2693

Schmuth M, Neyer S, Rainer C, Grassegger A, Fritsch P, Romani N, Heufler C (2002) Expression of the C-C chemokine MIP-3α/CCL20 in human epidermis with impaired permeability barrier function. Exp Dermatol 11:135–142

Schön MP, Ruzicka T (2001) Psoriasis: the plot thickens... Nat Immunol 2:91

Schröder JM, Mochizuki M (1999) The role of chemokines in cutaneous allergic inflammation. Biol Chem 389:889–896

Schwarz A, Beissert S, Grosse-Heitmeyer K, Gunzer M, Bluestone JA, Grabbe S, Schwarz T (2000) Evidence for functional relevance of CTLA-4 in ultraviolet-radiation-induced tolerance. J Immunol 165:1824–1831

Sebastiani S, Allavena P, Albanesi C, Nasorri F, Bianchi G, Traidl C, Sozzani S, Girolomoni G, Cavani A (2001) Chemokine receptor expression and function in CD4$^+$ T lymphocytes with regulatory activity. J Immunol 116:996–1002

Sebastiani S, Albanesi C, Nasorri F, Girolomoni G, Cavani A (2002) Nickel-specific CD8$^+$ and CD4$^+$ T cells display distinct migratory response to chemokines produced during allergic contact dermatitis. J Invest Dermatol 118:1052–1058

Sirsjo A, Karlsson M, Gidlof A, Rollman O, Torma H (1996) Increased expression of inducible nitric oxide synthase in psoriatic skin and cytokine-stimulated cultured keratinocytes. Br J Dermatol 134:643–648

Soumelis V, Reche PA, Kanzler H, Yuan W, Edward G, Homey B, Gilliet M, Ho S, Antonenko S, Lauerma A, Smith K, Gorman D, Zurawski S, Abrams J, Menon S, McClanahan T, de Waal-Malefyt Rd R, Bazan F, Kastelein RA, Liu YJ (2002) Human epithelial cells trigger dendritic cell-mediated allergic inflammation by producing TSLP. Nat Immunol 3:673–680

Sozzani S, Allavena P, Vecchi A, Mantovani A (2000) Differential regulation of chemokine receptors during dendritic cell maturation: a model for their trafficking properties. J Clin Immunol 20:151–160

Stingl G, Maurer D (1997) IgE-mediated allergen presentation via Fc epsilon RI on antigen-presenting cells. Int Arch Allergy Immunol 113:24–29

Stoof TJ, Flier J, Sampat S, Nieboer C, Tensen CP, Boorsma DM (2001) The antipsoriatic drug dimethylfumarate strongly suppresses chemokine production in human keratinocytes and peripheral blood mononuclear cells. Br J Dermatol 144:1114–1120

Taha RA, Minshall EM, Leung DY, Boguniewicz M, Luster A, Muro S, Toda M, Hamid QA (2000) Evidence for increased expression of eotaxin and monocyte chemotactic protein-4 in atopic dermatitis. J Allergy Clin Immunol 105:1002–1007

Traidl C, Sebastiani S, Albanesi C, Merk HF, Puddu P, Girolomoni G, Cavani A (2000) Disparate cytotoxic activity of nickel-specific CD8$^+$ and CD4$^+$ T cell subsets against keratinocytes. J Immunol 165:3058–3064

Trautmann A, Akdis M, Kleemann D, Altznauer F, Simon HU, Graeve T, Noll M, Brocker EB, Blaser K, Akdis CA (2000) T cell-mediated Fas-induced keratinocyte apoptosis plays a key pathogenetic role in eczematous dermatitis. J Clin Invest 106:25–35

Tsuji RI, Kawikova I, Ramabhadran R, Akahira-Azuma M, Taub D, Hugli T, Gerard C, Askenase PW (2000) Early local generation of C5a initiates the elicitation of contact sensitivity by leading to early T cell recruitment. J Immunol 165:1588–1598

Uguccioni M, Mackay CR, Ochensberger B, Loetscher P, Rhis S, LaRosa GJ, Rao P, Ponath PD, Baggiolini M, Dahinden CA (1997) High expression of the chemokine receptor CCR3 in human blood basophils. Role in activation by eotaxin, MCP-4, and other chemokines. J Clin Invest 100:1137–1143

Vanbervliet B, Homey B, Durand I, Massacrier C, Ait-Yahia S, de Bouteiller O, Vicari A, Caux C (2002) Sequential involvement of CCR2 and CCR6 ligands for immature dendritic cell recruitment: possible role at inflamed epithelial surfaces. Eur J Immunol 32:231–242

Varona R, Villares R, Carramolino L, Goya I, Zaballos A, Gutierréz J, Torres M, Martinez-A C, Màrquez G (2001) CCR6-deficient mice have impaired leukocyte homeostasis and altered contact hypersensitivity and delayed-type hypersensitivity responses. J Clin Invest 107:R37-R45

Vestergaard C, Bang K, Gesser B, Yoneymana H, Matsushima K, Larsen CG (2000) A Th2 chemokine, TARC, produced by keratinocytes may recruit CLA$^+$CCR4$^+$ lymphocytes into lesional atopic dermatitis skin. J Invest Dermatol 115:640–646

Vissers JLM, Hartgers FC, Lindhout E, Teunnisen MBM, Figdor CG, Adema GJ (2001) Quantitative analysis of chemokine expression by dendritic cell subsets in vitro and in vivo. J Leukoc Biol 69:785–793

von Andrian UH, Mackay CR (2000) T-cell function and migration. Two sides of the same coin. N Engl J Med 343:1020–1034.

von Bubnoff D, Geiger E, Bieber T (2001) Antigen-presenting cells in allergy. J Allergy Clin Immunol 108:329–339

Vulcano M, Albanesi C, Stopacciaro A, Bagnati R, D'Amico G, Struyf S, Transidico P, Bonecchi R, Del Prete A, Allavena P, Ruco LP, Chiabrando C, Girolomoni G, Mantovani A, Sozzani S (2001) Dendritic cells as a major source of macrophage-derived chemokine/CCL22 in vitro and in vivo. Eur J Immunol 31:812–822

Wakugawa M, Nakamura K, Kakinuma T, Onai N, Matsushima K, Tamaki K (2001) CC chemokine receptor 4 expression on peripheral blood CD4$^+$ T cells reflects disease activity of atopic dermatitis. J Invest Dermatol 117:188–196

Wang B, Fujisawa H, Zhuang L, Freed I, Howell BG, Shahid S, Shivji GM, Mak TW, Sauder DN (2000) CD4$^+$ Th1 and CD8$^+$ type 1 cytotoxic T cells both play a crucial role in the full development of contact hypersensitivity. J Immunol 165:6783–6790

Weng Y, Siciliano SJ, Waldburger KE, Sirotina-Meisher A, Staruch MJ, Daugherty BL, Gould SL, Springer MS, DeMartino JA (1998) Binding and functional properties of recombinant and endogenous CXCR3 chemokine receptors. J Biol Chem 273:18288–18291

Yamamoto J, Adachi Y, Onoue Y, Adachi YS, Okabe Y, Itazawa T, Toyoda M, Seki T, Morohashi M, Matsushima K, Miyawaki T (2000) Differential expression of the chemokine receptors by the Th1- and Th2-type effector populations within circulating CD4$^+$ T cells. J Leukoc Biol 68:568–574

Yasukawa H, Sasaki A, Yoshimura A (2000) Negative regulation of cytokine signaling. Annu Rev Immunol 18:143–164

Yawalkar N, Uguccioni M, Scharer J, Braunwalder J, Karlen S, Dewald B, Braathen LR, Baggiolini M (1999) Enhanced expression of eotaxin and CCR3 in atopic dermatitis. J Invest Dermatol 113:43–48

Yokozeki H, Ghoreishi M, Takagawa S, Takayama K, Satoh T, Katayama I, Takeda K, Akira S, Nishiota K (2000) Signal transducer and activator of transcription 6 is essential in the induction of contact hypersensitivity. J Exp Med 191:995–1004

Zabel BA, Agace WW, Campbell JJ, Heath HM, Parent D, Roberts AI, Ebert EC, Kassam N, Qin S, Zovko M, LaRosa GJ, Yang LL, Soler D, Butcher EC, Ponath PD, Parker CM, Andrew DP (1999) Human G protein-coupled receptor GPR9–6/CC chemokine receptor 9 is selectively expressed on intestinal homing T lymphocytes, mucosal lymphocytes, and thymocytes and is required for thymus-expressed chemokine-mediated chemotaxis. J Exp Med 190:1241–1256

Zlotnik A, Yoshie O (2000) Chemokines: a new classification system and their role in immunity. Immunity 12:121–127

Subject Index

Ernst Schering Research Foundation Workshop

Editors: Günter Stock
Monika Lessl

Vol. 1 (1991): Bioscience ⇋ Societly Workshop Report
Editors: D. J. Roy, B. E. Wynne, R. W. Old

Vol. 2 (1991): Round Table Discussion on Bioscience ⇋ Society
Editor: J. J. Cherfas

Vol. 3 (1991): Excitatory Amino Acids and Second Messenger Systems
Editors: V. I. Teichberg, L. Turski

Vol. 4 (1992): Spermatogenesis – Fertilization – Contraception
Editors: E. Nieschlag, U.-F. Habenicht

Vol. 5 (1992): Sex Steroids and the Cardiovascular System
Editors: P. Ramwell, G. Rubanyi, E. Schillinger

Vol. 6 (1993): Transgenic Animals as Model Systems for Human Diseases
Editors: E. F. Wagner, F. Theuring

Vol. 7 (1993): Basic Mechanisms Controlling Term and Preterm Birth
Editors: K. Chwalisz, R. E. Garfield

Vol. 8 (1994): Health Care 2010
Editors: C. Bezold, K. Knabner

Vol. 9 (1994): Sex Steroids and Bone
Editors: R. Ziegler, J. Pfeilschifter, M. Bräutigam

Vol. 10 (1994): Nongenotoxic Carcinogenesis
Editors: A. Cockburn, L. Smith

Vol. 11 (1994): Cell Culture in Pharmaceutical Research
Editors: N. E. Fusenig, H. Graf

Vol. 12 (1994): Interactions Between Adjuvants, Agrochemical
and Target Organisms
Editors: P. J. Holloway, R. T. Rees, D. Stock

Vol. 13 (1994): Assessment of the Use of Single Cytochrome
P450 Enzymes in Drug Research
Editors: M. R. Waterman, M. Hildebrand

Vol. 14 (1995): Apoptosis in Hormone-Dependent Cancers
Editors: M. Tenniswood, H. Michna

Vol. 15 (1995): Computer Aided Drug Design in Industrial Research
Editors: E. C. Herrmann, R. Franke

Supplement 1 (1994): Molecular and Cellular Endocrinology of the Testis
Editors: G. Verhoeven, U.-F. Habenicht

Supplement 2 (1997): Signal Transduction in Testicular Cells
Editors: V. Hansson, F. O. Levy, K. Taskén

Supplement 3 (1998): Testicular Function:
From Gene Expression to Genetic Manipulation
Editors: M. Stefanini, C. Boitani, M. Galdieri, R. Geremia,
F. Palombi

Supplement 4 (2000): Hormone Replacement Therapy
and Osteoporosis
Editors: J. Kato, H. Minaguchi, Y. Nishino

Supplement 5 (1999): Interferon: The Dawn of Recombinant
Protein Drugs
Editors: J. Lindenmann, W. D. Schleuning

Supplement 6 (2000): Testis, Epididymis and Technologies
in the Year 2000
Editors: B. Jégou, C. Pineau, J. Saez

Supplement 7 (2001): New Concepts in Pathology and
Treatment of Autoimmune Disorders
Editors: P. Pozzilli, C. Pozzilli, J.-F. Kapp

Supplement 8 (2001): New Pharmacological Approaches
to Reproductive Health and Healthy Ageing
Editors: W.-K. Raff, M. F. Fathalla, F. Saad

Supplement 9 (2002): Testicular Tangrams
Editors: F. F. G. Rommerts, K. J. Teerds

Supplement 10 (2002): Die Architektur des Lebens
Editors: G. Stock, M. Lessl

This series will be available on request from
Ernst Schering Research Foundation, 13342 Berlin, Germany

Printing: Saladruck Berlin
Binding: Stürtz AG, Würzburg